Contributions to Statistics

For further volumes:
http://www.springer.com/series/2912

Matilde Bini · Paola Monari · Domenico Piccolo ·
Luigi Salmaso
Editors

Statistical Methods for the Evaluation of Educational Services and Quality of Products

Physica-Verlag

Editors
Prof. Matilde Bini
University of Florence
Dept. of Statistics "G. Parenti"
Viale Morgagni, 59
50134 Firenze
Italy
bini@ds.unifi.it

Prof. Paola Monari
University of Bologna
Dept. Statistical Science
Via Belle Arti, 41
40126 Bologna
Italy
paola.monari@unibo.it

Prof. Domenico Piccolo
University of Naples Federico II
Dept. of Statistical Science
Via Leopoldo Rodinò, 22
80138 Napoli
Italy
domenico.piccolo@unina.it

Prof. Luigi Salmaso
University of Padua
Department of Management and Engineering
Stradella San Nicola, 3
36100 Vicenza
Italy
luigi.salmaso@unipd.it

ISBN 978-3-7908-2384-4   e-ISBN 978-3-7908-2385-1
DOI 10.1007/978-3-7908-2385-1
Springer Heidelberg Dordrecht London New York

Library of Congress Control Number: 2009939931

© Springer-Verlag Berlin Heidelberg 2009
This work is subject to copyright. All rights are reserved, whether the whole or part of the material is concerned, specifically the rights of translation, reprinting, reuse of illustrations, recitation, broadcasting, reproduction on microfilm or in any other way, and storage in data banks. Duplication of this publication or parts thereof is permitted only under the provisions of the German Copyright Law of September 9, 1965, in its current version, and permission for use must always be obtained from Springer. Violations are liable to prosecution under the German Copyright Law.

The use of general descriptive names, registered names, trademarks, etc. in this publication does not imply, even in the absence of a specific statement, that such names are exempt from the relevant protective laws and regulations and therefore free for general use.

Cover design: Integra Software Services Pvt. Ltd., Pondicherry

Printed on acid-free paper

Physica-Verlag is a brand of Springer-Verlag Berlin Heidelberg
Springer is part of Springer Science+Business Media (www.springer.com)

Preface

The book presents statistical methods and models that can usefully support the evaluation of educational services and quality of products. The contributions collected in this book summarize the work of several researchers from the universities of Bologna, Firenze, Napoli and Padova. The contributions are written with a consistent notation and a unified view, and concern methodological advances developed mostly with reference to specific problems of evaluation using real data sets.

The evaluation of educational services, as well as the analysis of judgements and preferences, poses severe methodological challenges because of the presence of one or more of the following aspects: the observational (non experimental) nature of the context, which is associated with the well-known problems of selection bias and presence of nuisance factors; the hierarchical structure of the data, that entails correlated observations and consideration of effects at different levels of the hierarchy and their interactions (multilevel analysis); the multivariate and qualitative nature of the dependent variable, that requires the use of ad hoc statistical methodologies; the presence of non observable factors, e.g. the satisfaction, calling for the use of latent variables models; the simultaneous presence of components of pleasure and components of uncertainty in the explication of the judgments, that asks for the specification and estimation of mixture models.

The first part of the book deals with latent variable models. In many fields of application most of the variables under investigation are not directly observable, and hence not measurable. In this context latent variable models assume a prominent role. Traditionally, latent variable models were used in psychometrics and have been concerned with measurement error, and latent variable constructs measured with multiple indicators (factor analysis). Nowadays, latent variables are used to represent different phenomena, such as true variables measured with error, hierarchical and longitudinal data, unobserved heterogeneity and missing data. Chapters 2, 3 and 4 illustrate latent variable models with educational behaviour applications. Since the variables under investigation are abilities, initial status, or rate of change in temporal achievement, the models rely on continuous latent variables, but different types of observations can be considered. Latent variable models for hierarchical data, i.e. multilevel models, are considered in Chaps. 5 and 6. In particular,

Chapter 5 reviews the use of multilevel models for value-added analysis in education. Chapter 6 describes the specification and estimation of a multilevel mixture factor model with continuous and categorical latent variables.

From a different point of view, Chap. 7 proposes an approach mainly based on individual perceptions about the discrete choices. In this framework, the latent process guiding the preferences and the judgements is represented by a mixture model. Extensions dealing with multi-attribute methods, such as conjoint analysis and choice modelling, are provided in Chap. 8, carrying out a brief and critical review in order to clarify the distinctions between the models as well as to point out their common issues.

A frequently encountered problem in fitting statistical models is the presence of outliers. Chapter 9 deals with a robust diagnostic approach known as Forward Search that detects the presence of outliers and assesses their influence on the estimates of the model parameters. In particular, the use of this approach is investigated in generalized liner models applied in studies on university performance evaluation.

The last chapters are devoted to nonparametric hypotheses testing via permutation methods for complex observational studies and to nonparametric construction of composite indicators. Chapter 10 presents a novel global performance score for the construction of a global performance index when the focus is at evaluating the product performances in connection with more than one aspect (dimension) and/or under several conditions (strata). Chapter 11 considers permutation methods for multivariate testing on ordered categorical variables within the framework of multivariate randomised complete block designs with application to a case study related to food sensorial evaluation. Chapter 12 is devoted to permutation tests for stochastic ordering problems where the main goal is to find out where the treatment peak is located (so called "umbrella alternative"). Chapter 13 deals with a novel method for constructing preference rankings based on the nonparametric combination procedure with application to the evaluation of professional profiles of municipal directors.

The Editors would like to thank all the people who, by their intensive research and aptitude of integration, have contributed to the realization of this book.

We thank Carla Rampichini of University of Florence for her precious collaboration to the editing work.

Matilde Bini - University of Firenze
Paola Monari - University of Bologna
Domenico Piccolo - University of Napoli Federico II
Luigi Salmaso - University of Padova

The Research Units were partially supported by a research grant from the Italian Ministry of University and Research (MIUR): PRIN 2006 "Statistical Methods and Models for the Evaluation of the Educational Processes", by the University of Padova CPDA088513 and by the CFEPSR, Portici.

Contents

1 Introduction 1
Matilde Bini, Paola Monari, Domenico Piccolo and Luigi Salmaso
1.1 Generalized linear latent variable models 1
1.2 Multilevel models 5
  1.2.1 Multilevel mixture factor models 6
1.3 Choices and conjoint analysis: critical aspects and recent developments 7
1.4 Robust diagnostic analysis with forward search 8
1.5 Nonparametric combination of dependent permutation tests and rankings 10
  1.5.1 Introduction to permutation tests 11
  1.5.2 Multivariate permutation tests and nonparametric combination methodology 12
  1.5.3 Nonparametric combination of dependent rankings 14

2 Latent variable models for ordinal data 17
Silvia Cagnone, Stefania Mignani and Irini Moustaki
2.1 Introduction 17
2.2 The GLLVM for ordinal data 18
  2.2.1 Model specification 18
  2.2.2 Model estimation 19
2.3 The goodness-of-fit of the model 21
  2.3.1 The problem of sparseness 21
  2.3.2 An overall goodness-of-fit test 23
2.4 GLLVM for longitudinal ordinal data 24
2.5 Case study: perceptions of prejudice on American campus 25
2.6 Concluding remarks 28

3 Issues on item response theory modelling 29
Mariagiulia Matteucci, Stefania Mignani and Bernard P. Veldkamp
3.1 Introduction 29

3.2	Basics of item response theory		30
	3.2.1	Parameter estimation	32
3.3	Advances in IRT: some issues		36
	3.3.1	Multidimensionality	37
	3.3.2	Incomplete design	40
	3.3.3	Inclusion of prior information	41
3.4	Case study: prior information in educational assessment		43
3.5	Concluding remarks		44

4 Nonlinearity in the analysis of longitudinal data 47
Estela Bee Dagum, Silvia Bianconcini and Paola Monari

4.1	Introduction		47
4.2	Latent curve models		48
	4.2.1	Estimation	50
4.3	Modelling nonlinearity in latent curve models		50
	4.3.1	Polynomial trajectories	51
	4.3.2	Exponential trajectories	52
	4.3.3	Complex nonlinear curves	53
4.4	Case study: analysis of university student achievements		55
	4.4.1	The data	55
	4.4.2	Results	57
4.5	Concluding remarks		60

5 Multilevel models for the evaluation of educational institutions: a review 61
Leonardo Grilli and Carla Rampichini

5.1	The evaluation of educational institutions		61
5.2	Effectiveness		62
	5.2.1	The value-added approach	64
	5.2.2	Type A and type B effectiveness	64
5.3	Multilevel models as a tool for measuring effectiveness		65
	5.3.1	The random intercept model	66
	5.3.2	The random slope model	69
	5.3.3	Cross-level interactions	69
	5.3.4	Fixed versus random effects	70
	5.3.5	Non-linear and multivariate multilevel models	71
	5.3.6	Multilevel models for non-hierarchical structures	72
5.4	Issues in model specification		72
	5.4.1	Simple versus complex models	72
	5.4.2	To adjust or not to adjust?	73
	5.4.3	Endogeneity	74
	5.4.4	Modelling the achievement progress	75
	5.4.5	Measurement error	76
5.5	Use of the model results		77
	5.5.1	Ranking the schools	77

		5.5.2 Predicting the outcome	79
	5.6	Concluding remarks	79

6	**Multilevel mixture factor models for the evaluation of educational programs' effectiveness** 81
	Roberta Varriale and Caterina Giusti
	6.1 Introduction 81
	6.2 The multilevel mixture factor model 82
	6.3 The Generalized Latent Variable framework 85
	6.4 Likelihood, estimation and posterior analysis 87
	6.5 Model selection 90
	6.6 Case study 91

7	**A class of statistical models for evaluating services and performances** 99
	Marcella Corduas, Maria Iannario and Domenico Piccolo
	7.1 Introduction 99
	7.2 Unobserved components in the evaluation process 100
	7.2.1 Rationale for a new class of models 101
	7.3 Specification and properties of CUB models 102
	7.4 Inferential issues and numerical procedures 106
	7.4.1 The EM algorithm 106
	7.4.2 Fitting measures 109
	7.5 Fields of application 110
	7.6 Further developments: a clustering approach 111
	7.7 Case study 112
	7.8 Concluding remarks 117

8	**Choices and conjoint analysis: critical aspects and recent developments** 119
	Rossella Berni and Riccardo Rivello
	8.1 Introduction 119
	8.2 Literature review 120
	8.2.1 Choice Experiment: theory and advances 123
	8.2.2 Conjoint Analysis: theory and advances 128
	8.3 Our proposal: conjoint analysis and response surface methodology 130
	8.3.1 The outlined theory 130
	8.3.2 The searching of the best profile through optimization ... 132
	8.4 Case study 133
	8.4.1 Optimization results 134
	8.5 Concluding remarks 137

9	**Robust diagnostics in university performance studies** 139
	Matilde Bini, Bruno Bertaccini and Silvia Bacci
	9.1 Introduction 139
	9.2 Robust methods vs diagnostic analysis 142

		9.2.1	The Forward Search algorithm 145
	9.3	\multicolumn{2}{l}{The Forward Search for Generalized Linear Models .... 146}

    9.3   The Forward Search for Generalized Linear Models 146
           9.3.1   Robust GLMs for the university effectiveness evaluation. The case of the first year college drop out rate 147
    9.4   The Forward Search for ANOVA models...................... 153
           9.4.1   The Forward Search for the fixed effects ANOVA....... 154
           9.4.2   The Forward Search for the random effects ANOVA 156
           9.4.3   The use of the robust ANOVA for the evaluation of the Italian university reform 158
    9.5   Concluding remarks 159

10 A novel global performance score with an application to the evaluation of new detergents 161
Stefano Bonnini, Livio Corain, Antonio Cordellina, Anna Crestana, Remigio Musci and Luigi Salmaso
    10.1 Introduction 161
    10.2 Composite indexes 163
          10.2.1 Standardization: data transformations to obtain homogeneous variables............................. 163
          10.2.2 Aggregation: synthesis of information 165
    10.3 Global performance score 167
          10.3.1 Global score on primary performance................. 167
          10.3.2 Global score on secondary performance............... 169
          10.3.3 Aggregation of GSP and GSS 170
    10.4 Case study: comparative performance evaluations of new detergents... 170
    10.5 A comparative simulation study 173
    10.6 Conclusions 176

11 Nonparametric tests for the randomized complete block design with ordered categorical variables 181
Livio Corain and Luigi Salmaso
    11.1 Introduction 181
    11.2 Overview on procedures proposed in the literature for the RCB design 182
    11.3 Permutation tests for multivariate RCB design 185
    11.4 Simulation study 188
    11.5 Case study 190
    11.6 Conclusions 192

12 A permutation test for umbrella alternatives 193
Dario Basso, Fortunato Pesarin and Luigi Salmaso
    12.1 Introduction 193
    12.2 Simple stochastic ordering alternatives 196
    12.3 Permutation test for umbrella alternatives 199
    12.4 Simulation study 201
    12.5 Case study: graduates in engineering 204

13 Nonparametric methods for measuring concordance between rankings: a case study on the evaluation of professional profiles of municipal directors ... 209
Rosa Arboretti Giancristofaro, Mario Bolzan and Livio Corain
- 13.1 Introduction ... 209
- 13.2 Sample survey of municipal directors' professional profiles ... 210
  - 13.2.1 Context of the evaluation of the role of Communes and of municipal directors ... 210
  - 13.2.2 Sample survey among Communes of the Veneto ... 211
- 13.3 Analysis of concordance between rankings ... 212
  - 13.3.1 The construction of ranks ... 212
  - 13.3.2 Hypothesis testing on concordance between rankings ... 213
  - 13.3.3 Closed testing procedure ... 216
  - 13.3.4 Rankings of the municipal director's professional profile in the Communes of the Veneto survey ... 219
  - 13.3.5 Discussion ... 220

References ... 227

Index ... 241

Contributors

Silvia Bacci
Department of Statistics "G. Parenti", University of Florence, Viale Morgagni 59, 50134, Florence, Italy, s.bacci@ds.unifi.it

Dario Basso
Department of Management and Engineering, University of Padua Stradella S. Nicola 4, 36100, Vicenza, Italy, basso@gest.unipd.it

Rossella Berni
Department of Statistics "G. Parenti", University of Florence, Viale Morgagni 59, 50134, Florence, Italy, berni@ds.unifi.it

Bruno Bertaccini
Department of Statistics "G. Parenti", University of Florence, Viale Morgagni 59, 50134, Florence, Italy, brunob@ds.unifi.it

Silvia Bianconcini
Department of Statistical Sciences, University of Bologna, via Belle Arti 41, 40126 Bologna, Italy, silvia.bianconcini@unibo.it

Matilde Bini
Department of Statistics "G. Parenti", University of Florence, Viale Morgagni 59, 50134, Florence, Italy, bini@ds.unifi.it

Mario Bolzan
Department of Statistics, University of Padua, Via Cesare Battisti 241, 35121- Padua, Italy, mbolzan@stat.unpd.it

Stefano Bonnini
Department of Mathematics, University of Ferrara, Via Machiavelli 35, 44100 Ferrara, Italy, bnnsfn@unife.it

Silvia Cagnone
Department of Statistical Sciences, University of Bologna, via Belle Arti 41, 40126 Bologna, Italy, silvia.cagnone@unibo.it

Livio Corain
Department of Management and Engineering, University of Padua, Stradella
S.Nicola 3, 36100 Vicenza, Italy, livio.corain@unipd.it

Antonio Cordellina
Research & Development, Reckitt-Benckiser Ltd, Piazza S. Nicolo' 12/3, 30034
Mira (VE), Italy, Antonio.Cordellina@Reckittbenckiser.com

Marcella Corduas
Department of Statistical Sciences, University of Naples Federico II, Via Leopoldo
Rodinò, 22 I-80138, Naples, Italy, marcella.corduas@unina.it

Anna Crestana
Research & Development, Reckitt-Benckiser Ltd, Piazza S. Nicolo' 12/3, 30034
Mira (VE), Italy, Anna.Crestana@Reckittbenckiser.com

Estela Bee Dagum
Department of Statistical Sciences, University of Bologna, via Belle Arti 41, 40126
Bologna, Italy, estela.beedagum@unibo.it

Rosa Arboretti Giancristofaro
Department of Territory and Agro-Forestal Systems, University of Padua, Viale
dell'Univeristà 16, 35020 Legnaro (PD), Italy, rosa.arboretti@unipd.it

Caterina Giusti
Department of Statistics and Mathematics Applied to Economics, University of
Pisa, Via C. Ridolfi 10, 56124, Pisa, Italy, caterina.giusti@ec.unipi.it

Leonardo Grilli
Department of Statistics "G. Parenti", University of Florence, Viale Morgagni 59,
50134, Florence, Italy, grilli@ds.unifi.it

Maria Iannario
Department of Statistical Sciences, University of Naples Federico II, Via Leopoldo
Rodinò, 22 I-80138, Naples, Italy, maria.iannario@unina.it

Mariagiulia Matteucci
Department of Statistical Sciences, University of Bologna, Via Belle Arti 41, 40126
Bologna, Italy, m.matteucci@unibo.it

Stefania Mignani
Department of Statistical Sciences, University of Bologna, via Belle Arti 41, 40126
Bologna, Italy, stefania.mignani@unibo.it

Paola Monari
Department of Statistical Sciences, University of Bologna, via Belle Arti 41, 40126
Bologna, Italy, paola.monari@unibo.it

Irini Moustaki
Department of Statistical Science, London School of Economics, Houghton Street,
London, WC2A 2AE UK, i.moustaki@lse.ac.uk

Remigio Musci
Research & Development, Reckitt-Benckiser Ltd, Piazza S. Nicolo' 12/3, 30034 Mira (VE), Italy, Remigio.Musci@Reckittbenckiser.com

Fortunato Pesarin
Department of Statistics, University of Padua Via Cesare Battisti 241, 35121, Padua, Italy,
pesarin@stat.unipd.it

Domenico Piccolo
Department of Statistical Sciences, University of Naples Federico II, Via Leopoldo Rodinò, 22 I-80138, Naples, Italy, domenico.piccolo@unina.it

Carla Rampichini
Department of Statistics "G. Parenti", University of Florence, Viale Morgagni 59, 50134, Florence, Italy, rampichini@ds.unifi.it

Riccardo Rivello
Department of Statistics "G. Parenti", University of Florence, Viale Morgagni 59, 50134, Florence, Italy, rivello@ds.unifi.it

Luigi Salmaso
Department of Management and Engineering, University of Padua, Stradella S.Nicola 3, 36100 Vicenza, Italy, salmaso@gest.unipd.it

Roberta Varriale
Department of Methodology and Statistics, Tilburg University, Tilburg, The Netherlands; Department of Statistics "G. Parenti", University of Florence, Florence, Italy, roberta.varriale@ds.unifi.it

Bernard P Veldkamp
Department of Research Methodology, Measurement and Data Analysis, University of Twente, P.O. Box 217, 7500 AE Enschede, The Netherlands,
b.p.veldkamp@gw.utwente.nl

Chapter 1
Introduction

Matilde Bini, Paola Monari, Domenico Piccolo and Luigi Salmaso

1.1 Generalized linear latent variable models

In many fields of application most of the variables under investigation are not directly observable and hence not measurable. In these contexts latent variable models assume a prominent role. Their origins can be traced back to the early twentieth century, notably in the study of human abilities. The main ideas lie behind factor analysis and the newer applications of linear structural models. An account of their innovative role in many fields to which statistical methods are applied can be found in Bartholomew (1995) and Bartholomew and Knott (1999). In the recent literature there have been several proposals for generalized latent variable modelling frameworks, integrating specific methodologies in a global theoretical context. One example is the Generalized Linear Latent And Mixed Models (GLLAMM) framework of Skrondal and Rabe-Hesketh (2004). This approach unifies and extends latent variable modelling as multilevel, longitudinal, and structural equation models as well as generalized linear mixed models, random coefficient models, item response models, factor models, and so on. Other two examples are Muthén (2008) and Vermunt (2007), both proposing general frameworks that allow to define models with any

---

Matilde Bini
Department of Statistics "G. Parenti", University of Florence, Viale Morgagni 59, 50134, Florence, Italy, e-mail: bini@ds.unifi.it

Paola Monari
Department of Statistical Sciences, University of Bologna, via Belle Arti 41, 40126 Bologna, Italy, e-mail: paola.monari@unibo.it

Domenico Piccolo
Department of Statistical Sciences, University of Naples Federico II, Via Leopoldo Rodinó, 22 I-80138, Napoli, Italy, e-mail: domenico.piccolo@unina.it

Luigi Salmaso
Department of Management and Engineering, University of Padua, Stradella S. Nicola 3, 36100 Vicenza, Italy, e-mail: salmaso@gest.unipd.it

combination of categorical and continuous latent variables at each level of the hierarchy.

Latent variable models specify the joint distribution of a set of observed and latent variables. Variables that are directly observed, also known as manifest variables, will be denoted by Y. A collection of K manifest variables will be distinguished by subscripts and written as column vector $\mathbf{y} = (Y_1, ..., Y_K)'$. Latent variables will be denoted by X, and Q latent variables will form the column vector $\boldsymbol{\eta}$. In practice, Q will be much smaller than K. Both latent and manifest variables can be metrical and/or categorical and vary from one individual to another. The relationships between them must be expressed in terms of probability distributions, so that, after the Y's have been observed, the information we have about $\boldsymbol{\eta}$ is given by its conditional distribution given $\mathbf{y}$

$$h(\boldsymbol{\eta}|\mathbf{y}) = \frac{h(\boldsymbol{\eta})g(\mathbf{y}|\boldsymbol{\eta})}{f(\mathbf{y})} \qquad (1.1)$$

where $h(\boldsymbol{\eta})$ is the prior distribution of $\boldsymbol{\eta}$, and $g(\mathbf{y}|\boldsymbol{\eta})$ is the conditional distribution of $\mathbf{y}$ given $\boldsymbol{\eta}$. As only $\mathbf{y}$ can be observed, any inference must be based on the joint distribution whose density may be expressed as

$$f(\mathbf{y}) = \int_{\mathbf{R}_\eta} h(\boldsymbol{\eta})g(\mathbf{y}|\boldsymbol{\eta})d\boldsymbol{\eta}, \qquad (1.2)$$

where $\mathbf{R}_\eta$ is the range space of $\boldsymbol{\eta}$.

The main assumption in this framework is the *conditional (or local) independence* of the observations $\mathbf{y}$ given the latent variables $\boldsymbol{\eta}$. Hence, Q must be chosen so that

$$g(\mathbf{y}|\boldsymbol{\eta}) = \prod_{i=1}^{K} g_i(y_i|\boldsymbol{\eta}) \qquad (1.3)$$

A latent variable model consists of two parts. The first part is given by the prior distribution of the latent variables $h(\boldsymbol{\eta})$. This accounts for the nature of $\boldsymbol{\eta}$, but it was seen to be essentially arbitrary and its choice is largely a matter of convention. The second element in the model is the set of conditional distributions of the manifest variables given the latent variables $g_i(y_i|\boldsymbol{\eta})$. A convenient family of distributions which allows to account for both discrete and continuous observations is the *exponential family*

$$g_i(y_i|\boldsymbol{\eta}) = \exp\left\{\frac{y\theta - b(\theta_i)}{\phi_i} + d(y, \phi_i)\right\} \qquad (1.4)$$

where θ_i is some function of $\boldsymbol{\eta}$. The simplest assumption about the form of this function is to suppose that it is a linear function, in which case we have

$$\theta_i = \alpha_{i0} + \alpha_{i1}\eta_{i1} + ... + \alpha_{iQ}\eta_{iQ} \qquad i = 1, 2, ..., K \qquad (1.5)$$

This is the General Linear Latent Variable Model (GLLVM). The term "linear" refers to its linearity in the αs.

Several statistical methodologies based on observed and latent variables of different nature are encompassed in the GLLVM described above and they are formalized by Bartholomew and Knott (1999). It provides a generalization of both the classical Generalized Linear Models (GLMs) by including latent dependent variables, and the classical factor model by allowing observations of different nature as well as linear relationships among the factors. From this point of view, GLLVM also generalizes the LISREL model by describing the relationship between dependent and independent latent variables in terms of probability distributions.

Chapters 2, 3 and 4 illustrate GLLVM with application in the educational evaluation. Since the variables under investigation are abilities, initial status and rate of change in temporal achievement, we deal with continuous latent variables, but different types of observations are considered. From a different point of view, Chapter 7 proposes an approach mainly based on individual perceptions about the discrete choice; thus, latent variables are a fundamental issues but they are quantified by explicit parameters in the model and by subjects covariates when it is convenient.

Chapter 2 deals with the problem of ordinal observations. In the literature (Jöreskog and Moustaki, 2001) there are two main approaches for conducting latent variable analysis with categorical observed data. The most popular is the Underlying Variable Approach (UVA) which assumes that each manifest variable is an indirect observation of a standardized normal variable. This approach is used in the general framework of structural equation modelling (LISREL). The other main approach is the Item Response Function (IRF) approach by which the manifest variables are treated as they are. The unit of analysis is the entire response pattern of a subject, so no loss of information occurs. The models for ordinal data within the IRF has been recently developed by Moustaki (2000). After a review of basic concepts of the two approaches, some methodological developments are introduced. This methodological extension requires an improvement of the computational algorithms for parameter estimation. Furthermore some theoretical results on the goodness of fit problems due to the severe sparseness, typical of variables with many categories, are presented.

Chapter 3 deals with Item Response Theory (IRT), or latent trait models for the study of individual responses to a set of items designed to measure latent abilities. IRT is a measurement theory that was first formalized in the Sixties with the fundamental work of Lord and Novick (1968) and it has a predominant role in educational testing. An IRT model describes the relationship between the observable examinee performance in the test, typically in the form of responses to categorical items, and the unobservable latent ability. Therefore, IRT models can be included in the GLLVM framework. IRT is used in all phases of test administration, from the test calibration to the estimation of individual abilities, in which the estimated item parameters are used to characterize the examinees. After a brief presentation of the main assumptions of IRT models, several aspects related to specific problems in the context of test administration are treated. This decision has been motivated by the many advances introduced over the last few years, that allow both to support more complex models and to improve the estimation algorithms. In particular, issues on multidimensionality (Wang et al., 2004), incomplete design (Béguin and Glas, 2001) and the inclusion of prior information (van der Linden, 1999) are discussed, referring both to current literature and to some contributions of the authors. A particular attention is given to the use of the Gibbs sampler, in the Markov Chain Monte Carlo (MCMC) methods, for the estimation of IRT models (Albert, 1992; Fox and Glas, 2001). Finally, applications related to these topics are presented in the context of educational assessment.

Chapter 4 describes the application of GLLVM for the analysis of individual data repeated over a period of time, that allows dynamical studies of social processes, rather than static cross-sectional analyses. The analysis of repeated measures has been considered from different points of view, such as individual growth techniques (Singer and Willett, 2004), time series and econometric analysis (Diggle et al., 1994), and multilevel modelling (Skrondal and Rabe-Hesketh, 2004). They can be encompassed into the general class of random coefficient models, in which random effects are incorporated into the model in view of reflecting unobserved heterogeneity in the individual behavior. More generally, the random coefficients can be incorporated into GLLVM by considering them as latent variables. Borrowing from Meredith and Tisak (1990), we refer to these models as Latent Curve Models (LCMs), since random coefficients permit each case in the sample to have a different trajectory over time. Growth curve models are studied to compare University student careers over time. We focus on continuous response variables, using conventional normal-theory estimators, such as maximum likelihood, into the framework of GLLVM.

In Chapter 7 we assume that evaluation is a psychological process where a rater/judge expresses the agreement within a prefixed scale. This process is generated by the perception of value/quality/performance and is governed by latent variables. In order to model the empirical results of a survey and to infer on the stochastic mechanism that generated ordinal data, we suppose that the final choice is determined by personal feeling/attractiveness towards the item and intrinsic uncertainty always present in human decisions. These aspects are combined in an effective way by introducing a mixture random variable where both components are expressed and weighted, as in D'Elia and Piccolo (2005). Thus, we will introduce CUB models by considering the observed ordinal response y as a realization of a discrete random variable Y defined on the support $\{y = 1, 2, \ldots, m\}$, for a given integer $m > 3$, as a mixture of Uniform and Shifted Binomial random variables. Formally, its probability mass function is defined by:

$$\Pr(Y = y) = \pi \binom{m-1}{y-1} (1-\xi)^{y-1} \xi^{m-y} + (1-\pi) \frac{1}{m}, \quad y = 1, 2, \ldots, m,$$

where $\pi \in (0, 1]$ and $\xi \in [0, 1]$. By examining the π parameter we quantify the *propensity* of the respondent to adhere to a completely random choice whereas $1 - \xi$ parameter is related to the strength of feeling. Recently, Iannario (2009c) proved that these models are fully identifiable. This probability structure adheres to most of observed shapes for real ordinal data and it has been generalized to take into account the effect of significant covariates (Piccolo and D'Elia, 2008) or atypical situations (Iannario, 2009b). Then, asymptotic maximum likelihood inference has been developed (Piccolo, 2006) by using EM algorithm and a software in R is currently available (Iannario and Piccolo 2009) for the estimation of CUB models, without and with covariates. In this regard, a few application to real data set concerning university evaluation of teaching and services will be discussed. A special topic is a model-based clustering procedure, firstly performed by Corduas (2008a,c), where a Kullback-Liebler divergence criterion is applied for selecting subgroups of expressed ratings by university students. The characteristic of the proposal is the possibility to assess classical classification methods by an inferential approach within the unique framework of ordinal modelling. Although CUB models are focused on the marginal distribution of the respondents, their use seems effective for investigating sound relationships among ordinal responses and covariates and for enhancing unobserved traits in the data. Thus, differences and integrations with IRT are worth of interest.

1.2 Multilevel models

The class of multilevel models is suitable for the analysis of hierarchical data, where level 1 units are nested in level 2 units, which are possibly nested in level 3 units and so on. For example, students nested in classrooms, classrooms nested in schools, schools nested in districts. Longitudinal and repeated measures data can be seen as special cases of hierarchical data, with occasions nested in subjects.

The basic two-level model is the linear random intercept model:

$$y_{ij} = \alpha + \beta x_{ij} + \gamma w_j + u_j + e_{ij} \qquad (1.6)$$

where j indexes the level 2 units (clusters) and i indexes the level 1 units (subjects). For example, in the evaluation of schools the clusters are the schools and the subjects are the students. The variables in the model are:
- y_{ij} the outcome of subject i of cluster j;
- x_{ij} a vector with the features of subject i of cluster j;
- w_j a vector with the features of cluster j.

Then, u_j is the random effect of cluster j, i.e. an unobservable quantity characterizing such a cluster and shared by all its subjects. The term u_j is a residual component that captures all the relevant factors at the cluster level not accounted for by the covariates and thus its meaning depends on which covariates enter the model. The effect u_j is called random because it is a random variable, assuming independence among the clusters. For consistency of the estimates, the crucial assumption on u_j is that its expectation conditionally on the covariates is null (exogeneity). Less crucial, but standard assumptions are the homoscedasticity, i.e. u_j has constant variance, and the normality of the distribution. Finally, the level 1 errors e_{ij} are residual components taking into account all the unobserved factors at the subject level making the outcome different from what predicted by the covariates and the random effect. The e_{ij} are assumed independent among subjects and independent of u_j. The other standard assumptions are similar to those on u_j, i.e. exogeneity, homoscedasticity and normality. Model (1.6) is named random intercept since each cluster has its own intercept that has both fixed and random components. However, the slopes are assumed to be constant across clusters, so the regression lines are parallel.

The simple random intercept model (1.6) can be extended in several ways. For example, it is often found that the relationship between the outcome y_{ij} and a level 1 covariate x_{ij} varies from cluster to cluster, so the regression lines are no longer parallel. This leads to the so called random coefficient model that can be written as

$$y_{ij} = \alpha + u_{0j} + (\beta + u_{1j})x_{ij} + e_{ij} \qquad (1.7)$$

where it is usually assumed that (u_{0j}, u_{1j}) is bivariate normal. The random coefficient also implies that the between-cluster variance is a quadratic function of the covariate.

Now there are plenty of textbooks on multilevel modelling. Snijders and Bosker (1999) is an excellent introduction. Hox (2002) has fewer details, but it covers a wider range of topics. Raudenbush and Bryk (2002) present the models in a careful way along with thoroughly discussed applications. Goldstein (2003) is a classical, though not easy, reference with wide coverage and many educational applications.

Chapter 5 deals with the use of multilevel models for value-added analysis in education. The chapter reviews the concept of effectiveness in the educational setting and outlines the value-added approach. Multilevel models are presented as a tool for measuring effectiveness, with a discussion of several issues in model specification, such as the choice of the set of the covariates and the modelling of the achievement progress. The chapter ends with some remarks on the use of the model results for ranking the schools and for predicting the outcome of an hypothetical student.

1.2.1 Multilevel mixture factor models

Factor analysis is a well-known statistical method used to describe the correlations among some manifest variables, indicators, in terms of fewer latent variables, factors. In its standard formulation, factor analysis assumes that the variables are measured on a set of independent units; this assumption may be inadequate when units are nested in clusters assuming what is called a hierarchical structure (Goldstein, 2003; Snijders and Bosker, 1999). These differences can be modeled including group dummies in the model, as in the multigroup approach, or can be modeled with a multilevel factor model with continuous latent variables at all level of the analysis. Besides the difference in the nature, fixed or random, of the group effects, these two models differ in their perspective: the multilevel factor model usually aims at exploring the latent structure underlying the observed phenomenon at different levels of the analysis (see, as some examples, Goldstein and McDonald (1988), Longford and Muthén (1992) and Grilli and Rampichini (2007a)) while the multigroup factor model has a confirmatory approach and aims at comparing the observed groups of units with respect to the different parameters of the factor model (Bollen, 1989; Meredith, 1993; Muthén, 1989). In a confirmatory perspective, another model useful to compare the observed groups of units is the multilevel mixture factor model with a categorical latent variable at the higher level of the analysis. This model evaluates the existence of unobserved subpopulation (classes) of groups with similar features with respect to the factor model parameters and overcomes the creation of over-detailed information of the multigroup factor model, which estimates as many group coefficients as the groups. Mixture factor analyses have been developed and largely used in the one-level context (McLachlan and Peel, 2000; Magidson and Vermunt, 2001; Lubke and Muthén, 2005). More recently, the specification of mixture factor models in the multilevel context has received a growing interest. As en example, Palardy and Vermunt (2009) and Muthén (2008) use a two-level mixture model in the context of growth analysis and Vermunt (2007) use a mixture model in the context of IRT analysis; Muthén and Asparouhov (2008) also describe a more general two-level mixture model with different types of latent variables.

Chapter 6 deals with the use of multilevel models in the context of factor analysis and, more precisely, in the context of mixture factor models. This chapter describes the specification and estimation of a multilevel mixture factor model with continuous latent variables at the lower level of the analysis and a categorical latent variable at the higher level focusing, on one hand, on the illustration of some theoretical issues of the model and, on the other hand, on the applied results that can be achieved Varriale (2007). Then, a multilevel mixture factor model is used in order to evaluate the external effectiveness of the Italian university using, as indicator of the phenomenon, the information on the job satisfaction expressed by the graduates. In particular, the model is used to analyse the underlying structure of the job satisfaction at the individual level and, at the same time, to cluster higher level units represented by the programs that the individuals attended during the university in classes with some typical characteristics.

1.3 Choices and conjoint analysis: critical aspects and recent developments

Standard conjoint analysis (CA) is a multi-attribute quantitative method useful to study the evaluation of a consumer/user about a new product/service. In the literature many authors (see for example Alvarez-Farizo and Hanley (2002)), have studied and applied this method; the main theoretical problems are faced by Green and Srinivasan (1990) about statistical models and by Moore (1980), related to the insertion of baseline variables related to the respondent.

In Chapter 8 a joint study including a modified conjoint analysis and the Response Surface Methodology, in order to improve the analysis of multi-attribute valuation methods, is presented.

Our proposal is based on the conjoint analysis jointly with the status-quo evaluation, Hartman et al. (1991), which is the alternative related to the current situation.

The statistical analysis is carried out through the Response Surface Methodology (RSM) (for more details see Khuri and Cornell (1987) and Myers and Montgomery (1995)) by considering the quantitative judgement of each respondent for each profile with respect to the assessed score about the status-quo and taking the individual information into account. The final result is achieved carrying out an optimization procedure on the estimated statistical models, by defining an objective function in order to reach the optimal solution for the revised (or new) service/product. In this context, it is relevant to point out the modified structured data, through a new questionnaire, in order to collect information about the baseline variables of the respondent, the quantitative data about the current situation (status-quo) of the product/service and the proper CA analysis by means of the planning of an experimental design.

In general, we may define the set of experimental variables, which influence the measurement process: $\mathbf{x} = [\mathbf{x}_1,..,\mathbf{x}_k,..,\mathbf{x}_K]$; and the set of noise variables: $\mathbf{z} = [\mathbf{z}_1,..,\mathbf{z}_s,..,\mathbf{z}_S]$.

The general RSM model can be written as:

$$\mathbf{Y}_{ij}(\mathbf{x},\mathbf{z}) = \beta_0 + \mathbf{x}'\beta + \mathbf{x}'\mathbf{B}\mathbf{x} + \mathbf{z}'\delta + \mathbf{z}'\Delta\mathbf{z} + \mathbf{x}'\Lambda\mathbf{z} + \mathbf{e}_{ij} \quad i=1,..,I; j=1,..,J \quad (1.8)$$

where $\mathbf{x}$ and $\mathbf{z}$ are the vectors of variables as described above; β, $\mathbf{B}$, δ, Δ, and Λ are vectors and matrices of the model parameters, $\mathbf{e}_{ij}$ is the random error which is assumed Normally distributed with zero mean and variance equal to σ. Λ is a $K \times S$ matrix which plays an important role since it contains the parameters of the interaction effects between the $\mathbf{x}$ and $\mathbf{z}$ sets.

Note that, in general, if J are the profiles and I the respondents, the observations are $I \times J$. In this context, the set $\mathbf{x}$ are the judgements, expressed through votes in a metric scale [0,100], on the attributes involved in the experimental planning; while the set $\mathbf{z}$ is related to the baseline individual variables, which are relevant for the service or product studied and that may change according to the specific situation. The response variable $\mathbf{Y}$ is defined as a quantitative variable of the process; in this case, the judgements expressed, on each full profile of the plan, by the respondents in the same metric scale.

The final aim is to find the best preference, by evaluating both the quantitative judgements about the full profiles and the judgements about the current situation, which is the most insensitive to the heterogeneity of the respondent.

1.4 Robust diagnostic analysis with forward search

A frequently encountered difficulty in statistical inference problems is the presence of outliers in the data. Outliers can be defined as observations which appear to be inconsistent or somewhat different from the rest of the data. They can arise from models different from the one we intend to estimate (contaminants) or can be

1 Introduction

atypical observations generated by the assumed model. Their identification is of extreme importance since they can have strong negative effects on classical estimator efficiency, and should hence be eliminated or down-weighted in the estimation of the model. Furthermore, their pattern should be thoroughly examined since they could provide valuable new information on the problem being analyzed. Unfortunately, their identification is often very difficult, particularly when multivariate distributions are being dealt with.

The Forward Search, introduced by Atkinson and Riani (2000), is a general diagnostic approach for detecting the presence of outliers and assessing their influence on the estimates of the model parameters. The method was applied to regression analysis, but it could as well be applied to almost any model and multivariate method (Atkinson et al., 2004).

This algorithm is based on the following steps: the start is a robust fit to very few observations and then a successive fit is done with larger subsets. More specifically, it starts by finding a presumably outlier-free subset of observations, for example the set proposed by Rousseeuw (1984) to find the least median of squares estimators (LMS), i.e., the value of the parameters that yields the smallest median squared residual. The surface to be minimized has many local minima and the minimum value can only be obtained by approximation. Rousseeuw proposed restricting the search to all the estimates obtained by using only subsets of size p. The starting subset of the Forward Search, is given by the p observations which yield the smallest median squared residual. This is an approximation of the real LMS estimate and unfortunately still requires the evaluation of all possible subsets of size p (Bertaccini and Bini, 2007).

Formally, let $Z = (X, y)$ a data matrix of dimension $nx(p+1)$. If n is moderate and $p << n$, the choice of the initial subset can be performed by exhaustive enumeration of all $\binom{n}{p}$ distinct ptuple $S^{(p)}_{i_1,...,i_p} \equiv \{z_1,...,z_p\}$, where Z^T_{ij} is the i_jth row of Z, for and $1 < i_j \neq i_{j^*} < n$. Specifically, let $\iota^T = [i_1,...,i_p]$ and let $e_{\iota, S^{(p)}_t}$ be the least squares residual for the unit i given the model has been fitted with the observations in $S^{(p)}_t$. The initial subset is which satisfies $e^2_{[med], S^{(p)}_*} = \min_t \left[e^2_{[med], S^{(p)}_t} \right]$ where $e^2_{[k], S^{(p)}_*}$ is the kth ordered squared residual among $e^2_{[i], S^{(p)}_*}$, with $i = 1,...,n$ and med = integer part of $(n+p+1)/2$. If $\binom{n}{p}$ is too large, the choice is made using 3,000 ptuples sampled from Z matrix.

The subset size is increased by one and the model refitted to the observations with the smallest residuals for the increased subset size.

The initial subset $S^{(m)}_*$ of dimension $m \geq p$ is increased by one and the new subset $S^{(*)}_{m+1}$ consists of $m*1$ units with the smallest ordered residuals $e^2_{[k], S^{(m)}_*}$. The model is refitted to the new subset and the procedure continues with increasing subset sizes until all the data are fitted, i.e. when $S^{(m)}_* = S^{(n)}$. The result is an ordering of the observations by their closeness to the assumed model. Usually one observation enters the subset used for fitting, but sometimes two or more observations enter the subset as one or more leave.

Chapter 9 proposes to validate this robust diagnostic approach when university performance analyses are carried out. In particular, the algorithm is investigated in generalized liner models. The analysis also reviews some robust studies recently performed about effectiveness and efficiency of Italian universities (Bini et al., 2002; Biggeri and Bini, 2003); Bini et al., 2003; Bini and Bertaccini, 2004; Bini, 2004a, 2004b; Bertaccini and Polverini, 2006).

1.5 Nonparametric combination of dependent permutation tests and rankings

Chapters 10, 11, 12 and 13 of the book deal with the Nonparametric Combination approach of dependent permutation Tests (NPC Test) and Rankings (NPC Ranking) to face a variety of univariate and multivariate problems for the evaluation of educational services and quality of products. After a short abstract of each chapter, in this section we provide an introduction on notation and basic theory of nonparametric combination methodology of permutation tests or rankings.

Chapter 10 presents a novel Global Performance Score (GPS) for the construction of a global performance index when we are facing a complex problem of product quality evaluation, that is when the focus is on evaluating the product performances in connection with more than one aspect (dimension) and/or under several conditions (strata). The methodological solution we propose to cope with this problem is described and applied, considering different possible data transformation and an application problem related to the performance evaluation of new detergents.

Chapter 11 considers permutation methods for testing on ordered categorical variables within the framework of randomised complete block designs. The proposed approach is studied and validated via a Monte Carlo simulation study and it has been applied to a food sensorial evaluation study.

Chapter 12 is devoted to permutation tests for stochastic ordering problems where the main goal is to find out where the treatment peak is located (so called "umbrella alternative"). The proposed solution involves testing for stochastic ordering of continuous variables and the nonparametric combination methodology. Since the location of the peak is generally unknown, it can be detected by sequential tests on possible picks and combining together those tests.

Chapter 13 deals with a novel method for constructing preference rankings based on the nonparametric combination procedure and the proposed method is compared with that based on the arithmetic mean. Subsequently, in order to verify to what extent two rankings concord, a new permutation test for the evaluation of concordance between dependent rankings is developed. Finally, the method is applied to the evaluation of professional profiles of municipal directors.

1.5.1 Introduction to permutation tests

The importance of the permutation approach in resolving a large number of inferential problems is well-documented in the literature, where the relevant theoretical aspects emerge, as well as the extreme effectiveness and flexibility from an applicatory point of view (Manly, 1997; Pesarin, 2001; Edgington and Onghena, 2007; Basso et al., 2009).

The great majority of univariate problems may be usefully and effectively solved within standard parametric or nonparametric methods as well, although in relatively mild conditions their permutation counterparts are generally asymptotically as good as the best parametric ones. Moreover, it should be noted that permutation methods are essentially of a nonparametrically exact nature in a conditional context. In addition, there are a number of parametric tests the distributional behavior of which is only known asymptotically. Thus, for most sample sizes of practical interest, the relative lack of efficiency of permutation solutions may sometimes be compensated by the lack of approximation of parametric asymptotic counterparts. In addition, assumptions regarding the validity of parametric methods (such as normality and random sampling) are rarely satisfied in practice, so that consequent inferences, when not improper, are necessarily approximated, and their approximations are often difficult to assess.

For any general testing problem, in the null hypothesis (H_0), which usually assumes that data come from only one (with respect to groups) unknown population distribution P, the whole set of observed data $\mathbf{x}$ is considered to be a random sample, taking values on sample space $\mathscr{X}^n$, where $\mathbf{x}$ is one observation of the n-dimensional sampling variable $\mathbf{X}^{(n)}$ and where this random sample does not necessarily have independent and identically distributed (i.i.d.) components. We note that the observed data set $\mathbf{x}$ is always a set of sufficient statistics in H_0 for any underlying distribution.

Given a sample point $\mathbf{x}$, if $\mathbf{x}^* \in \mathscr{X}^n$ is such that the likelihood ratio $f_P^{(n)}(\mathbf{x})/f_P^{(n)}(\mathbf{x}^*)$ $= \rho(\mathbf{x}, \mathbf{x}^*)$ is not dependent on f_P for whatever $P \in \mathscr{P}$, then $\mathbf{x}$ and $\mathbf{x}^*$ are said to *contain essentially the same amount of information* with respect to P, so that they are equivalent for inferential purposes. The set of points that are equivalent to $\mathbf{x}$, with respect to the information contained, is called the *orbit associated with* $\mathbf{x}$, and is denoted by $\mathscr{X}^n_{/\mathbf{x}}$, so that $\mathscr{X}^n_{/\mathbf{x}} = \{\mathbf{x}^* : \rho(\mathbf{x}, \mathbf{x}^*) \text{ is } f_P\text{-independent}\}$.

The same conclusion is obtained if $f_P^{(n)}(\mathbf{x})$ is assumed to be invariant with respect to permutations of the arguments of $\mathbf{x}$; i.e., the elements $(x_1, \ldots, x_n)$. This happens when the assumption of independence for observable data is replaced by that of *exchangeability*, $f_P^{(n)}(x_1, \ldots, x_n) = f_P^{(n)}\left(x_{u_1^*}, \ldots, x_{u_n^*}\right)$, where $(u_1^*, \ldots, u_n^*)$ is any permutation of $(1, \ldots, n)$. Note that, in the context of permutation tests, this concept of exchangeability is often referred to as the *exchangeability of the observed data with respect to groups*. Orbits $\mathscr{X}^n_{/\mathbf{x}}$ are also called *permutation sample spaces*. It is important to note that orbits $\mathscr{X}^n_{/\mathbf{x}}$ associated with data sets $\mathbf{x} \in \mathscr{X}^n$ always contain a finite number of points, as n is finite.

Since, in the null hypothesis and assuming exchangeability, the conditional probability distribution of a generic point $\mathbf{x}' \in \mathscr{X}^n_{/\mathbf{x}}$, for any underlying population distribution $P \in \mathscr{P}$, is P-independent, permutation inferences are invariant with respect to P in H_0. Some authors, emphasizing this invariance property, prefer to give them the name of *invariant tests*. However, due to this invariance property, permutation tests are distribution-free and nonparametric.

Formally, let $\mathscr{X}^n/\mathbf{x}$ be the orbit associated with the observed vector of data $\mathbf{x}$. The points of $\mathscr{X}^n/\mathbf{x}$ can also be defined as $\mathbf{x}^* : \mathbf{x}^* = \pi\mathbf{x}$ where π is a random permutation of indexes $1, 2, \ldots, n$. Define a suitable test statistic T on $\mathscr{X}^n/\mathbf{x}$ for which large values are significant for a right-handed one-sided alternative: The support of $\mathscr{X}^n/\mathbf{x}$ through T is the set $\mathscr{T}$ that consists of C elements (if there are no ties in the given data). Let

$$T^*_{(1)} \leq T^*_{(2)} \leq \cdots \leq T^*_{(C)}$$

be the ordered values of $\mathscr{T}$. Let T^o be the observed value of the test statistic, $T^o = T(\mathbf{x})$. For a chosen attainable significance level $\alpha \in \{1/C, 2/C, \ldots, (C-1)/C\}$, let $k = C(1-\alpha)$. Define a permutation test, the function $\phi^* = \phi(T^*)$ for a one-sided alternative

$$\phi^*(T) = \begin{cases} 1 & \text{if} \quad T^o \geq T^*_{(k)} \\ 0 & \text{if} \quad T^o < T^*_{(k)} \end{cases}.$$

Permutation tests have general good properties such as exactness, unbiasedness and consistency (see Pesarin, 2001; Hoeffding, 1952).

1.5.2 Multivariate permutation tests and nonparametric combination methodology

In this section, we provide details on the construction of multivariate permutation tests via nonparametric combination approach. Consider, for instance, a multivariate problem where q (possibly dependent) variables are considered. The main difficulties arise because of the underlying dependence structure among variables (or aspects), which is generally unknown. Moreover, a global answer involving several dependent variables (aspects) is often required, so the question is how to combine the information related to the q variables (aspects) into one global test.

In a multivariate problem, when the aim is to compare two o more groups, the matrix of data is generally partitioned into n q-dimensional arrays; that is,

$$\mathbf{X_{n \times q}} = \begin{bmatrix} x_{11} & x_{12} & \cdots & x_{1q} \\ x_{21} & x_{22} & \cdots & x_{2q} \\ \vdots & \vdots & \ddots & \vdots \\ x_{n1} & x_{n2} & \cdots & x_{nq} \end{bmatrix}.$$

1 Introduction

Each row of **X** is a determination of the multivariate variable $[X_1, X_2, \ldots, X_q]$, which has distribution P with unknown dependence structure.

In this framework the null hypothesis H_0, which states the equality in distribution of the multivariate distribution of the q variables in all groups, is supposed to be properly decomposed into q sub-hypotheses H_{0j} each appropriate for partial (univariate) aspects,

$$H_0 : \bigcap_{j=1}^{q} H_{0j}.$$

Hence, the *global* null hypothesis H_0 can be viewed as an intersection of *partial* null hypotheses H_{0j}. Under the global null hypothesis, the rows of **X** are exchangeable. We can thus define q *partial* test statistics. Let T_j, $j = 1, \ldots, q$, be a *partial* test statistic for the univariate hypothesis H_{0j} involving each of the q variables.

A desirable property of a multivariate test is that the global null hypothesis should be rejected whenever one of the partial null hypothesis is rejected. To this end, let us consider the rule *large is significant*, which means that the global test statistic should assume large values whenever at least one of its arguments leads to the rejection of at least one partial null hypothesis H_{0j}. Accordingly, the global test ψ^* should be based on a suitable combining function ψ that satisfies the following requirements:

1. A combining function ψ must be non-increasing in each argument:

$$\psi(\lambda_1, \ldots, \lambda_j, \ldots, \lambda_q) \geq \psi(\lambda_1, \ldots, \lambda'_j, \ldots, \lambda_q) \quad \text{if} \quad \lambda_j < \lambda'_j, \; j \in \{1, \ldots, q\}.$$

1. Every combining function ψ must attain its supremum value $\bar{\psi}$, possibly not finite, even when only one argument attains zero:

$$\psi(.., \lambda_j, ..) \to \bar{\psi} \quad \text{if} \quad \lambda_j \to 0, \; j \in \{1, \ldots, q\}.$$

2. $\forall \alpha > 0$, the critical value of every ψ is assumed to be finite and strictly smaller than the supremum value: $T''_\alpha < \bar{\psi}$.

The λ's in the definition of the combining function are p-values: $\alpha_i = \Pr\{T_i^* \geq T_i^o | H_{0i}\}$. It is possible, of course, to express ψ also in terms of partial statistics. For instance, if the λ's are test statistics that are significant for large values (as in the bivariate example), some suitable combining functions are the following:

- the direct combining function: $\psi = \sum_{j=1}^{q} \lambda_j$;
- the max$_T$ combining function: $\psi = \max_j \lambda_j$.

Instead, if the combining function is based on the partial p-values (i.e., $\lambda_j = p_j = \Pr\left[T_j^* \geq T_j | \mathbf{Y}\right]$, which are significant against H_{0j} for small values), the following combining functions are of interest:

- Fisher's: $\psi = -2 \sum_{j=1}^{q} \log(p_j)$, $0 \leq \psi \leq +\infty$;
- Tippett's: $\psi = 1 - \min_j p_j$, $0 \leq \psi \leq 1$;

- Liptak's: $\psi = \sum_{j=1}^{q} \phi^{-1}(1-p_j)$, where ϕ is the standard normal cumulative distribution function, $0 \leq \psi \leq +\infty$.

The global p-value is defined as:

$$p^G = \frac{1}{C} \sum_{b=1}^{C} I(\psi^* \geq \psi).$$

Remember that, in order to preserve the underlying dependence relations among variables, permutations must always be carried out on individual data vectors, so that all component variables and partial tests must be jointly analyzed. If the global p-value is significant, then there is empirical evidence that at least one partial null hypothesis is not true.

It can be proved that the multivariate permutation tests maintain the properties of univariate permutation tests (for details see Pesarin, 2001; Basso et al., 2009).

1.5.3 Nonparametric combination of dependent rankings

The main purpose of the nonparametric combination of dependent rankings method (NPC Ranking, Lago and Pesarin (2000)) is to obtain a single ranking indicator for the statistical units being studied, which summarizes many partial rankings. This method is defined as being nonparametric since it needs neither the knowledge of the underlying statistical distribution for the variables being studied, nor the dependence structure among partial rankings, apart for the assumption that all regressions are monotonic. Given a multivariate phenomenon $\mathbf{X} = [X_1, X_2, \ldots, X_q]$, observed on n statistical units, and once we have calculated the q partial rankings $R_1, R_2, \ldots, R_q$, starting from the variables X_j, $j = 1, \ldots, q$, each one being informative about a partial aspect of $\mathbf{X}$, we wish to build up a *global combined ranking* G:

$$G = \psi(X_1, X_2, \ldots, X_q; w_1, w_2, \ldots, w_q), \qquad \mathbb{R}^{2q} \to \mathbb{R},$$

where ψ is a real function allowing us to combine the partial dependent rankings and $w_1, w_2, \ldots, w_q$ is a set of weights, defined on the basis of technological, functional or economic considerations, which measure the relative degree of importance among the q aspects of $\mathbf{X}$.

In order to build up G we introduce a set of minimal reasonable conditions related to the variables X_j, $j = 1, \ldots, q$:

1. for each of the q informative variables a partial ordering criterion is well established, in the sense that "large is better" (if it is not so, it is possible to recode the variables by means of any appropriate transformation like $1/X$ or $-X$);
2. regression relationships within the q rankings are monotonic (increasing or decreasing).

1 Introduction

If by chance, the marginal distribution of any informative variable is degenerate, then the corresponding partial ranking can be discharged from analysis, since it is non informative. Indeed, in this case all n statistical units have assigned the same rank, hence the combined ranking is unaffected by it.

Moreover, we do not need any further assumption both on the statistical distribution of the informative variables, and on their dependence structure. Finally, notice that we do not need to assume the continuity of X_j, $j = 1,\ldots,q$, so that the probability of *ex-equo* can be positive.

Without loss of generality, X_j, $j = 1,\ldots,q$, are assumed to behave in accordance with the rule "large is better" and in this setting, we consider the rank transformations R_{ij} (*partial rankings*):

$$\{R_{ij} = R(X_{ij}) = \#(X_{ij} \geq X_{hj})\} \qquad i,h = 1,\ldots,n; j = 1,\ldots,q.$$

Associated with these ranks are the scores:

$$\left\{\lambda_{ij} = \frac{R_{ij} + 0.5}{n+1}, \ i = 1,\ldots,n; j = 1,\ldots,q\right\}.$$

Once a combining function ψ (for more details see Pesarin, 2001) has been chosen, we compute the transformation

$$\psi : \{Q_i = \psi(\lambda_{i1},\ldots,\lambda_{iq}; w_1, w_2,\ldots,w_q), \ i = 1,\ldots,n\},$$

and finally, applying the rank transformation, we obtain the global combined ranking G:

$$\{G_i = R(Q_i) = \#(Q_i \geq Q_h), \ i,h = 1,\ldots,n\}$$

In the global ranking G, each statistical units is ranked in a unique way, by taking into consideration the whole set of the q informative variables.

The real combining function ψ is chosen from class Ψ of combining functions satisfying the following minimal properties:

1. ψ must be continuous in all $2q$ arguments, in that small variations in any subset of arguments imply a small variation in the ψ-index;
2. ψ must be monotone non-decreasing with respect to each argument:

$$\psi(\ldots,X_i,\ldots; w_1,\ldots,w_q) \geq \psi(\ldots,X_i',\ldots; w_1,\ldots,w_q)$$

   if $0 < X_i' < X_i < 1$, $i = 1,\ldots,n$;
3. ψ must be symmetric with respect to permutations of the arguments, in that if, for instance, $u_1,\ldots,u_q$ is any permutation of $1,\ldots,q$ then:

$$\psi(X_{u_1},\ldots,X_{u_q}; w_{u_1},\ldots,w_{u_q}) = \psi(X_1,\ldots,X_q; w_1,\ldots,w_q).$$

Property 1 is obvious. Property 2 means that if, for instance, two subjects have exactly the same values for all Xs, except for the i-th, then the one with $X_i > X_i'$ must have assigned at least the same ψ-index. Property 3 states that any combining function ψ must be invariant with respect to the order in which informative variables are processed.

Some of the combining functions most often used are Fisher, Liptak or Tippett. If in the overall analysis it is of interest to give different weights to partial rankings, then by using appropriate weights opportunely fixed: $w_j \geq 0$, $j = 1,\ldots,q$, the combining function using the Fisher becomes:

$$\psi_F = -\sum_{j=1}^{q} w_j \log(1-\lambda_j).$$

NPC Ranking has been proved to be effective with respect to standard methods to combine simple indicators (Arboretti et al., 2008). Moreover, it has been successfully applied in the industrial field for development and quality assessment of new products (Bonnini et al., 2006; Corain and Salmaso, 2007).

Chapter 2
Latent variable models for ordinal data

Silvia Cagnone, Stefania Mignani and Irini Moustaki

2.1 Introduction

Latent variable models with observed ordinal variables are particularly useful for analyzing survey data. Typical ordinal variables express attitudinal statements with response alternatives like "strongly disagree", "disagree", "strongly agree" or "very dissatisfied", "dissatisfied", "satisfied" and "very satisfied".

In the literature, there are two main approaches for analyzing ordinal observed variables with latent variables. The most popular one is the *Underlying Variable Approach* (UVA) (Muthén, 1984; Jöreskog, 1990) which assumes that the observed variables are generated by underlying normally distributed continuous variables. This approach is used in structural equation modeling and the relevant methodological developments are available in commercial software such as LISREL (Jöreskog and Sörbom, 1988) and Mplus (Muthén and Muthén, 1998–2007). The other approach is the *Item Response Theory* (IRT) according to which the observed variables are treated as they are. The unit of analysis is the entire response pattern of a subject, so no loss of information occurs. An overview of those type of models can be found in Bartholomew and Knott (1999) and van der Linden and Hambleton (1997). Moustaki and Knott (2000) and Moustaki (2000) discuss a Generalized Linear Latent Variable Model framework (GLLVM) for fitting models with different types of observed variables.

Silvia Cagnone
Department of Statistical Sciences, University of Bologna, via Belle Arti 41, 40126 Bologna, Italy
e-mail: silvia.cagnone@unibo.it

Stefania Mignani
Department of Statistical Sciences, University of Bologna, via Belle Arti 41, 40126 Bologna, Italy
e-mail: stefania.mignani@unibo.it

Irini Moustaki
Department of Statistical Science, London School of Economics, Houghton Street, London, WC2A 2AE, UK, e-mail: i.moustaki@lse.ac.uk

Several studies (Jöreskog and Moustaki, 2001; Huber et al., 2004; Cagnone et al., 2004) showed that the latter approach is preferable in terms of accuracy of estimates and model fit. This is due to the fact the UVA is based on limited information estimation methods whereas IRT is a full information approach. However, full information methods are much more computationally intensive especially as the number of latent variables increases. Solutions to computational problems for IRT models have been recently proposed by Huber et al. (2004) and Schilling and Bock (2005).

In the following sections we review the latent variable models for ordinal data within the GLLVM framework as introduced by Moustaki (2000). The chapter will focus on the goodness-of-fit issue when sparseness is present (Reiser, 1996; Maydeu-Olivares and Harry, 2005; Cagnone and Mignani, 2007) and the most recent extension to longitudinal data (Cagnone et al., 2009). An application to a subset of the National Longitudinal Survey of Freshmen (NLSF) is also presented.

2.2 The GLLVM for ordinal data

2.2.1 Model specification

Let **y** be a vector of K ordinal observed variables each of them with c_k categories and $\boldsymbol{\eta}$ a vector of Q latent variables. The c_k ($k=1,\ldots,K$) ordered categories of the variables y_k have associated probabilities $\pi_{1,k}(\boldsymbol{\eta}), \pi_{2,k}(\boldsymbol{\eta}), \ldots, \pi_{c,k}(\boldsymbol{\eta})$ which are functions of the vector of the latent variables $\boldsymbol{\eta}$. Within this framework, the unit of analysis is the response pattern of an individual; for the r-th individual it is defined as $\mathbf{y}_r = (y_1 = s_1, y_2 = s_2, \ldots, y_K = s_K)$. There are NR $= \prod_{k=1}^{K} c_k$ possible response patterns.

The probability associated to $\mathbf{y}_r$ is given by

$$f(\mathbf{y}_r) = \pi_r = \int_{R_\eta} g(\mathbf{y}_r | \boldsymbol{\eta}) h(\boldsymbol{\eta}) d\boldsymbol{\eta} = \int_{R_\eta} \pi(\boldsymbol{\eta}) h(\boldsymbol{\eta}) d\boldsymbol{\eta} \quad (2.1)$$

where $h(\boldsymbol{\eta})$ is assumed to be a multivariate normal distribution with $\mathbf{0}$ mean and correlation (or covariance) matrix equal to $\boldsymbol{\Phi}$ and $g(\mathbf{y}_r | \boldsymbol{\eta})$ is the conditional probability of the observed variables given the latent variables following a multinomial distribution. Under the assumption of conditional independence:

$$g(\mathbf{y}_r | \boldsymbol{\eta}) = \prod_{k=1}^{K} g(y_k | \boldsymbol{\eta}) = \prod_{k=1}^{K} \pi_{s,k}^{y_{s,k}} = \prod_{k=1}^{K} (\gamma_{s,k} - \gamma_{s-1,k})^{y_{s,k}} \quad s = 2, \cdots, c_k \quad (2.2)$$

where $y_{s,k} = 1$ if a randomly selected individual responds into category s of the kth item and $y_{s,k} = 0$ otherwise. $\gamma_{s,k}$ is the cumulative probability of responding below category s. Unlike the model for binary data, in this case we define the conditional distribution $g(y_k | \boldsymbol{\eta})$ in terms of cumulative probabilities $\gamma_{s,k}$ since they take

2 Latent variable models for ordinal data

into account the ranking of the categories of the ordinal variables. In more detail $\gamma_{s,k} = \pi_{1,k} + \pi_{2,k} + \ldots + \pi_{s,k}$ is the probability of a response in category s or lower on the variable k. As in the classical generalized linear model, the relation between the observed and the latent variables can be expressed through any monotone differentiable link function. In the case of ordinal variables we can refer to the logit as follows:

$$\ln\left[\frac{\gamma_{s,k}}{(1-\gamma_{s,k})}\right] = \tau_{s,k} - \sum_{q=1}^{Q} \alpha_{kq}\eta_q, \quad s=1,\ldots,c_k-1 \qquad (2.3)$$

where $\tau_{s,k}$ and α_{kq} can be interpreted as thresholds and factor loadings of the model, respectively. The ordinality is defined properly by the condition $\tau_{1,k} \leq \tau_{2,k} \leq \ldots \leq \tau_{c-1,k}$. We refer to (2.3) as the Proportional Odds Model (POM) (McCullagh and Nelder, 1983).

2.2.2 Model estimation

The parameters of the model are estimated with the E-M algorithm. The E-M has been used for estimating the two-parameter logistic model for binary variables in Bock and Aitkin (1981), and then used for estimating the GLLVM in Bartholomew and Knott (1999). See Moustaki (2000) for the case of ordinal data.

Starting from Eq. (2.1) the joint density of the random variable for the ith individual can be written as

$$f(\mathbf{y}_i, \boldsymbol{\eta}_i) = g(\mathbf{y}_i|\boldsymbol{\eta}_i)h(\boldsymbol{\eta}_i). \qquad (2.4)$$

If we consider a sample of size n, the complete log-likelihood is given by:

$$\sum_{i=1}^{n} \log f(\mathbf{y}_i, \boldsymbol{\eta}_i) = \sum_{i=1}^{n} \log[g(\mathbf{y}_i|\boldsymbol{\eta}_i)h(\boldsymbol{\eta}_i)] = \sum_{i=1}^{n} [\log g(\mathbf{y}_i|\boldsymbol{\eta}_i) + \log h(\boldsymbol{\eta}_i)]. \qquad (2.5)$$

From the assumption of conditional independence we get:

$$\sum_{i=1}^{n} \log f(\mathbf{y}_i, \boldsymbol{\eta}) = \sum_{i=1}^{n}\left[\sum_{k=1}^{K} \log g(y_{ki}|\boldsymbol{\eta}_i) + \log h(\boldsymbol{\eta}_i)\right]. \qquad (2.6)$$

The thresholds and factor loadings are found in the first component of the log-likelihood whereas the parameters related with the covariance matrix of the latent variables are found in the second component.

Estimation of the correlation between latent variables. The E-M algorithm requires first the computation of the expected score function of the correlation terms with respect to the posterior distribution $h(\boldsymbol{\eta}|\mathbf{y})$

$$E_i S(\boldsymbol{\Phi}) = \int_{R_\eta} S(\boldsymbol{\Phi})h(\boldsymbol{\eta}|\mathbf{y}_i)d\boldsymbol{\eta} \qquad (2.7)$$

where
$$S(\boldsymbol{\Phi}) = \frac{\partial \log h(\boldsymbol{\eta}, \boldsymbol{\Phi})}{\partial \boldsymbol{\Phi}} \qquad (2.8)$$

that is
$$S(\boldsymbol{\Phi}) = \partial \log h(\boldsymbol{\eta}, \boldsymbol{\Phi})/\partial \boldsymbol{\Phi} = -\frac{1}{2}\boldsymbol{\Phi}^{-1} + \frac{1}{2}\boldsymbol{\Phi}^{-1}(\boldsymbol{\eta}\boldsymbol{\eta}')\boldsymbol{\Phi}^{-1}. \qquad (2.9)$$

By substituting (2.9) in Eq. (2.7) we get:
$$E_i S(\boldsymbol{\Phi}) = \int_{R_\eta} \left(-\frac{1}{2}\boldsymbol{\Phi}^{-1} + \frac{1}{2}\boldsymbol{\Phi}^{-1}(\boldsymbol{\eta}\boldsymbol{\eta}')\boldsymbol{\Phi}^{-1} \right) h(\boldsymbol{\eta}|\mathbf{y}_i) d\boldsymbol{\eta}. \qquad (2.10)$$

The integrals can be approximated by using the Gauss-Hermite quadrature points. Since the latent variables are correlated, the approximation is obtained by using the Choleski factorization of the correlation matrix $\boldsymbol{\Phi} = \mathbf{C}\mathbf{C}'$. The Gauss-Hermite approximation will be applied to the integral of the transformed variables as follows

$$f(\mathbf{y}) = (2\pi)^{-n/2} \sum_{w_1,\ldots,w_Q} g\left(\mathbf{z} \mid \mathbf{C}\left(\beta_{w_1},\ldots,\beta_{w_Q}\right)'\right) h\left(\mathbf{C}\left(\beta_{w_1},\ldots,\beta_{w_Q}\right)'\right) \qquad (2.11)$$

where $\boldsymbol{\eta} = \mathbf{C}\boldsymbol{\beta}$, $\sum_{w_1,\ldots,w_Q} = \sum_{w_1=1}^{v_1} \cdots \sum_{t_n=1}^{v_Q}$ and $v_1, \ldots, v_Q$ are the quadrature points. By solving $\sum_{i=1}^n E_i S(\boldsymbol{\Phi}) = 0$ using the above approximation, we get explicit solutions for the maximum likelihood estimator of the elements of $\boldsymbol{\Phi}$

$$[\hat{\boldsymbol{\Phi}}]_{lj} = \frac{\sum_{i=1}^n \sum_{w_1,\ldots,w_Q} \left[\left(\mathbf{C}\left(\beta_{w_1},\ldots,\beta_{w_Q}\right)'\right) \left(\mathbf{C}\left(\beta_{w_1},\ldots,\beta_{w_Q}\right)'\right)' \right]_{lj} h\left(\mathbf{C}\left(\beta_{w_1},\ldots,\beta_{w_Q}\right)|\mathbf{y}_i\right)}{\sum_{i=1}^n \sum_{w_1,\ldots,w_Q} h\left(\mathbf{C}\left(\beta_{w_1},\ldots,\beta_{w_Q}\right)|\mathbf{y}_i\right)}$$
(2.12)

Estimation of the parameters in $g(\mathbf{y}|\boldsymbol{\eta})$. The expected score function of the parameters $\mathbf{a}_k = \left(\tau_{1,k},\ldots,\tau_{c_k-1,k},\alpha_{k1},\ldots,\alpha_{kQ}\right), k=1,\ldots,K$ with respect of $h(\boldsymbol{\eta}|\mathbf{y})$ is given by

$$E_i S(\mathbf{a}_k) = \int_{R_\eta} S_i(\mathbf{a}_k) h(\boldsymbol{\eta}|\mathbf{y}_i) d\boldsymbol{\eta}, \qquad (2.13)$$

where in this case
$$S_i(\mathbf{a}_k) = \frac{\partial \log g(\mathbf{y}_i|\boldsymbol{\eta})}{\partial \mathbf{a}_k}. \qquad (2.14)$$

By solving $E_i S(\mathbf{a}_k) = 0$ we get not-explicit solutions for the parameters $\mathbf{a}_k$. The expressions of the derivatives (2.14) can be found in Moustaki (2000) and Moustaki (2003).

The E-M algorithm works as follows:

- Choose initial estimates for the model parameters.
- E-step: Compute the Expected score functions given in (2.7) and (2.13).

- M-step: Obtain improved estimates for the parameters by solving the non-linear maximum likelihood equations for the parameters of the conditional distribution $g(\mathbf{y}|\boldsymbol{\eta})$ by using a Newton-Raphson iterative scheme and explicit solutions for the correlations between the latent variables.
- Return to step 2 and continue until convergence is achieved.

2.3 The goodness-of-fit of the model

2.3.1 The problem of sparseness

The usual way of testing the goodness-of-fit of latent variable models for ordinal data is to compare the observed and the expected frequencies of all possible response patterns (NR). A test for the model may be based on the usual goodness-of-fit statistics such as the likelihood ratio (LR) and the Pearson chi-square test (GF), defined as follows:

$$\text{LR} = 2n \sum_{r=1}^{\text{NR}} f_r \ln\left(\frac{f_r}{\widehat{\pi}_r}\right), \quad (2.15)$$

$$\text{GF} = n \sum_{r=1}^{\text{NR}} \frac{(f_r - \widehat{\pi}_r)^2}{\widehat{\pi}_r}, \quad (2.16)$$

where f_r is the sample proportion of the r-th response pattern, $\widehat{\pi}_r$ is the corresponding estimated probability $\widehat{\pi}_r = \pi_r(\widehat{\mathbf{a}})$ and n is the sample size.

Under regular conditions both statistics are approximately distributed as a χ^2 with degrees of freedom $df = \text{NR} - 1 - \#\text{pr}$ where #pr is the number of the estimated parameters. With reference to the contingency table whose cells contain the frequencies of the response patterns, the number of observations in each cell should be large enough to justify the asymptotic approximation of the statistics to the chi-square distribution. Nevertheless, in many cases, contingency tables do not have large numbers of observations and the sparseness problem arise. To solve the sparseness problem a number of theoretical strategies has been proposed. Such strategies have been applied both to the goodness-of-fit statistics and to the residuals calculated from the marginal distributions of the observed variables. For a review of strategies applied to the former see Koheler and Larntz (1980), Agresti and Yang (1987), Read and Cressie (1988), Bartholomew and Tzamourani (1999), and Tollenar and Mooljaart (2003).

An alternative solution to the sparseness problem is to consider the residuals computed from marginal distributions. The residuals express the discrepancies between observed and expected frequencies and can be defined in a number of different ways. Residuals can provide information on how well the model predicts the one and two-way marginal distributions revealing items or pairs of items for which the model does not fit well. In fact, even in the presence of a severe degree of sparseness, almost always the univariate and the bivariate marginal frequencies distributions are quite large so that statistics based on these frequencies are not affected by sparseness. A thorough treatment of the analysis of residuals is given by Reiser (1996) with reference to the two-parameter item response model for binary data. The use of residuals in GLLVM for binary data is discussed in Bartholomew and Tzamourani (1999). They recommend to use them as supplementary analysis to the overall goodness-of-fit testing. In particular, they argue that a good model predicts well all the pairwise associations between observed variables. On the contrary, if some pairs of variables present high bivariate residuals, they indicate that the model does not fit the data. As for POM, Jöreskog and Moustaki (2001) have defined specific measures of fit based on the residuals. For the univariate marginal distributions they have proposed the following measure related to the GF (an equivalent measure is given also for the LR index but it is not reported here because it is outside the scope of this work):

$$\text{GF fit}^{(k)} = n \sum_{s=1}^{c_k} \frac{(f_{s,k} - \hat{\pi}_{s,k})^2}{\hat{\pi}_{s,k}} \quad k=1,\ldots,K \quad (2.17)$$

where we can define:

$$\hat{\pi}_{s,k} = \sum_{r=1}^{NR} y_{rs} \hat{\pi}_r, \quad (2.18)$$

and

$$y_{rs} = \begin{cases} 1 & \text{if } y_k = s \\ 0 & \text{otherwise.} \end{cases} \quad (2.19)$$

The quantities $(f_{s,k} - \hat{\pi}_{s,k})^2 / \hat{\pi}_{s,k}$ $(s=1,\ldots,c_k)$ are the standardized residuals computed from the univariate marginal distribution of the variable k.

In the same way, for the bivariate marginal distributions of the variables k and l we get:

$$\text{GF fit}^{(kl)} = n \sum_{sk,sl} \frac{(f_{sk,sl} - \hat{\pi}_{sk,sl})^2}{\hat{\pi}_{sk,sl}} \quad k=1,\ldots,K-1 \quad l=k+1,\ldots,K \quad (2.20)$$

where, as before, we can define:

$$\hat{\pi}_{sk,sl} = \sum_{r=1}^{NR} y_{rsk} y_{rsl} \hat{\pi}_r, \quad (2.21)$$

and

$$y_{rsk} = \begin{cases} 1 & \text{if } y_k = s_k \\ 0 & \text{otherwise,} \end{cases} \quad (2.22)$$

$$y_{rsl} = \begin{cases} 1 & \text{if } y_l = s_l \\ 0 & \text{otherwise.} \end{cases} \quad (2.23)$$

In this case the quantities $(f_{sk,sl} - \hat{\pi}_{sk,sl})^2 / \hat{\pi}_{sk,sl}$ $(s=1,\ldots,c_k; s=1,\ldots,c_l)$ are the standardized residuals computed from the bivariate marginal distribution of the variables k and l.

2.3.2 An overall goodness-of-fit test

The residuals based on the marginal distributions can be used for building a overall goodness-of-fit test. To this aim, we need to define the unstandardized residuals for the overall r-th response pattern as:

$$g_r = f_r - \hat{\pi}_r. \tag{2.24}$$

Under regular conditions (Birch, 1964), the NR dimensional vector $\sqrt{n}\mathbf{g}$ converges asymptotically to a gaussian random vector with mean equal to $\mathbf{0}$ and covariance matrix $\mathbf{\Omega_g}$ defined as:

$$\mathbf{\Omega_g} = \mathbf{D}(\boldsymbol{\pi}) - \boldsymbol{\pi}\boldsymbol{\pi}' - \mathbf{T}(\mathbf{F'F})^{-1}\mathbf{T'}, \tag{2.25}$$

where $\mathbf{D}(\boldsymbol{\pi})$ is a diagonal matrix that contains the NR probabilities π_r, the matrix $\mathbf{F}$ is defined as $\mathbf{F} = \mathbf{D}(\boldsymbol{\pi})^{-1/2}\partial\boldsymbol{\pi}/\partial\mathbf{a}$. Finally $\mathbf{T} = \partial\boldsymbol{\pi}/\partial\mathbf{a}$.

The residuals just defined are computed from the overall contingency table of the manifest variables. From these residuals it is possible to obtain the unstandardized residuals associated to the marginal distributions. We refer to the residuals for the bivariate marginal distributions (considering, for simplicity, only the case in which the observed variables have the same number of categories, that is $c_k = c_l = c$). For category a of variable k and category b of variable l they can be defined as:

$$e = (f_{sk,sl} - \hat{\pi}_{sk,sl}). \tag{2.26}$$

$\hat{\pi}_{sk,sl}$ is directly computed by the estimated response probabilities $\hat{\pi}_r$. Passing to the matrix form we can write:

$$\mathbf{e} = \mathbf{M}(\mathbf{f} - \hat{\boldsymbol{\pi}}) = \mathbf{Mg}, \tag{2.27}$$

where $\mathbf{M}$ is a matrix of 0s and 1s. The generic element of $\mathbf{M}$, $m_{sk,sl}$ is given by:

$$m_{sk,sl} = \begin{cases} 1 & \text{if } y_k = s \text{ and } y_l = s \\ 0 & \text{otherwise.} \end{cases} \tag{2.28}$$

The elements of $\mathbf{M}$ have been derived in such a way that multiplying $\mathbf{M}$ by the response probabilities $\boldsymbol{\pi}$, we realize the summation across the response patterns obtaining the second-order marginal proportions. From the asymptotic normality of $\mathbf{g}$ and from (2.27) we get:

$$\sqrt{n}\mathbf{e} \rightarrow N(\mathbf{0}, \mathbf{\Omega_e}) \tag{2.29}$$

where $\mathbf{\Omega_e} = \mathbf{M\Omega_g M'}$.
A consistent estimator for $\mathbf{\Omega_e}$ is given by:

$$\hat{\mathbf{\Sigma}}_e = n^{-1}\mathbf{M}(\mathbf{D}(\boldsymbol{\pi}) - \boldsymbol{\pi}\boldsymbol{\pi}' - \mathbf{T}(\mathbf{F'F})^{-1}\mathbf{T'})\mathbf{M'}|_{\boldsymbol{\alpha}=\hat{\boldsymbol{\alpha}},\boldsymbol{\pi}=\hat{\boldsymbol{\pi}}}. \tag{2.30}$$

The test of fit is developed for assessing the null hypothesis that the theoretical residuals are not significantly different from 0. With this regard we can refer to the

statistic:
$$X_e^2 = e' \hat{\Sigma}_e^+ e \quad (2.31)$$

that has an asymptotic χ^2 distribution. Since $\hat{\Sigma}_e$ is not a full rank matrix, its inversion can be obtained in different ways. Cagnone and Mignani (2007) propose to use the Moore-Penrose generalized inverse; in the case in which the computation of $\hat{\Sigma}_e^+$ is not stable, Maydeu-Olivares and Harry (2005) propose to compute a matrix that has $\hat{\Sigma}_e^+$ as generalized inverse. The degrees of freedom of the χ^2 depend on the rank of Σ_e, that in general results less or equal to the $\min\left(\sum_{k=0}^{2}\binom{p}{k}(c-1)^k, NR-1-(KQ+K(c-1))\right)$ namely, the minimum between the ranks of M and Ω_g (Bishop et al., 1975), respectively.

Reiser (1996) argued that, when sparseness is present, this index can be very useful for the goodness-of-fit of the overall model. In fact, although it is based on partial information, if higher-order interactions are not present (because of the conditional independence assumption) inferences regarding the parameters may be performed without loss of information in smaller marginal table (*collapsibility* of the contingency table). In this case this index produces good results in terms of both Type I error and power of the test (Reiser and Lin, 1999; Cagnone and Mignani, 2007). Nevertheless, when the collapsibility does not hold, this index is not as powerful as the indexes computed from the full contingency table.

2.4 GLLVM for longitudinal ordinal data

When questionnaires are submitted to the same individuals over time, we deal with longitudinal data or repeated measures. Recently many authors focused on latent variable models for longitudinal data with the aim of analyzing traits, attitudes, or any latent constructs over time (Roy and Lin, 2000; Dunson, 2003; Rabe-Hesketh et al., 1996). The latent variable model for ordinal data discussed in the previous sections has been extended to longitudinal data by Cagnone et al. (2009). The key feature of this model is that the inter-relationships among items are explained by time-dependent attitudinal latent variables whereas the associations across time are modelled via item-specific random effects. The time changes in the attitudinal latent variables are measured using a non-stationary autoregressive model. The resulted covariance matrix allows the latent variables to be correlated with unknown variances.

Formally, the model described in Sect. 2.2 is extended to longitudinal data in the following way. Given the vector of the K ordinal observed variables y_t measured at time t ($t = 1, \ldots, T$), the linear predictor defined in (2.3) becomes

$$\ln\left[\frac{\gamma_{t,k,s}}{(1-\gamma_{t,k,s})}\right] = \tau_{t,k,s} - \alpha_{kt}\eta_t - u_k, \quad k=1,\ldots,K; s_k=1,\ldots,c_k-1; t=1,\ldots,T$$
(2.32)

where the u_k's are item-specific random effects. The latent variables η_t and their variances allow to explain the associations among the items measured at time t. The associations among the same item measured across time are explained by u_k and the covariances between η_t's. The time dependent latent variables are related through a first order autoregressive structure

$$\eta_t = \phi \eta_{t-1} + \delta_t \qquad (2.33)$$

where for identification purposes $\delta_t \sim N(0,1)$ and $\eta_1 \sim N\left(0,\sigma_1^2\right)$. It is also assumed that the random effects u_k are independent of η_t and their common distribution function is $N_K(\mathbf{0}, \boldsymbol{\Sigma})$ with $\boldsymbol{\Sigma} = \mathrm{diag}_{k=1,\ldots,K}\left(\sigma_{uk}^2\right)$. It follows that $\mathrm{Var}(\eta_t) = \phi^{2(t-1)}\sigma_1^2 + I(t \geq 2)\sum_{l=1}^{t-1}\phi^{2(l-1)}$ and $\mathrm{Cov}(\eta_t, \eta_{t'}) = \phi^{t+t'-2}\sigma_1^2 + I(t \geq 2)\sum_{l=0}^{t-2}\phi^{t'-t+2l}$, where $I(.)$ is the indicator function.

As before, model estimation is obtained by using maximum likelihood estimation via the E-M algorithm. The substantial difference with the previous model in terms of estimation procedure is in the matrix $\boldsymbol{\Phi}$ whose elements express the relationships among both latent variables over time and latent variables and random effects. In more detail it is a covariance block matrix given by

$$\boldsymbol{\Phi} = \begin{bmatrix} \boldsymbol{\Gamma} & \mathbf{0} \\ \mathbf{0} & \boldsymbol{\Sigma} \end{bmatrix} \qquad (2.34)$$

where $\boldsymbol{\Gamma}$ is the variance covariance matrix of the time dependent latent variables. Its elements depend on the parameters ϕ and σ_1^2 in such a way that

$$\boldsymbol{\Gamma}^{-1} = \begin{bmatrix} \frac{1}{\sigma_1^2}+\phi^2 & -\phi & 0 & \ldots & 0 & 0 & 0 \\ -\phi & 1+\phi^2 & -\phi & \ldots & 0 & 0 & 0 \\ 0 & -\phi & 1+\phi^2 & \ldots & 0 & 0 & 0 \\ \vdots & \vdots & \vdots & \ddots & \vdots & \vdots & \vdots \\ 0 & 0 & 0 & \ldots & -\phi & 1+\phi^2 & -\phi \\ 0 & 0 & 0 & \ldots & 0 & -\phi & 1 \end{bmatrix}$$

Explicit solutions for the parameters ϕ, σ_1^2 and σ_{uk}^2 $(k=1,\ldots,K)$ are obtained whereas, as before, a Newton Raphson algorithm is used for the thresholds and the factor loadings of the model (Cagnone et al., 2009).

2.5 Case study: perceptions of prejudice on American campus

In order to illustrate the methodology described above we consider an example extracted from the National Longitudinal Survey of Freshmen (NLSF).[1] The NLSF

---

[1] This research is based on data from the National Longitudinal Survey of Freshmen, a project designed by Douglas S. Massey and Camille Z. Charles and funded by the Mellon Foundation and the Atlantic Philanthropies.

evaluates the academic and social progress of college students at regular intervals to capture emergent psychological processes, by measuring the degree of social integration and intellectual engagement and to control for pre-existing background differences with respect to social, economic, and demographic characteristics. Data are collected over a period of four waves (1999–2003). The sample was constituted by students of different races and 3,924 completed the survey.

In this analysis we concentrate on the part of questionnaire that investigates the perceptions of prejudice by the undergraduate students. It is composed by 13 ordinal items concerning different aspects of the perceptions of prejudice. After a preliminary exploratory factor analysis, we selected the following most important (in terms of reliability analysis) items:

1. How often, if ever, have students in your college classes ever made you feel uncomfortable or self-conscious because of your race or ethnicity? [StudUnc]
2. Walking around campus, how often, if ever, have you been made to feel uncomfortable or self-conscious because of your race or ethnicity? [CampUnc]
3. How often, if ever, have you felt you were given a bad grade by a professor because of your race or ethnicity [BadProf]
4. How often, if ever, have you felt you were discouraged by a professor from speaking out in class because of your race or ethnicity [DiscProf]

Permitted responses are "Never", "Rarely", "Sometimes", "Often", "Very often", "Don't know", "Refused". Since a small proportion of students responded to the last categories, categories from 3 to 5 have been collapsed leaving three categories for each item. Missing data have been treated by means of the listwise deletion. The final sample size is $n = 2,828$. The items are the same only for waves 2000 and 2001, hence in the analysis we consider two time points.

The aim of the analysis is first to fit at each time point a confirmatory factor model and then to perform a longitudinal analysis in order to evaluate if the perceptions of prejudice changes from 2000 to 2001. From a previous exploratory analysis we found that two factors can explain the variability between the items in both time points. Hence a POM model with correlated latent variables has been fitted to the data. In Table 2.1 the results of the estimates are reported.

Table 2.1 Parameter estimates with standard errors in brackets for the POM model, years 2000–2001, NLSF

Items	2000		2001	
	$\hat{\alpha}_{i1}$	$\hat{\alpha}_{12}$	$\hat{\alpha}_{i1}$	$\hat{\alpha}_{12}$
StudUnc	3.55 (0.32)	–	3.80 (0.23)	–
CampUnc	2.71 (0.18)	–	2.86 (0.14)	–
BadProf	–	2.88 (0.16)	–	4.36 (0.18)
DiscProf	–	3.79 (0.27)	–	2.73 (0.20)
ϕ_{12}	0.56(0.02)		0.62(0.01)	

We can observe that the loadings are high and significant for both factors and at both time points. Moreover they are very similar over time, indicating that the measurement invariance assumption is probably satisfied (same loadings over time, (Cagnone et al., 2009)). The correlations are quite high and significant in both observed years.

As for the goodness-of-fit, in year 2000 the LR and GF are equal to 188.52 and 190.58 respectively with $df = 51$ indicating that the two-factor model is rejected. The same result is obtained for year 2001, LR and GF being equal to 200.09 and 170.32 and $df = 51$. However, as discussed above, these tests can be affected by sparse data and therefore limited test statistics are computed instead, X_e^2. For 2000, we obtained $X_e^2 = 110.90$ with $df = 21$ and for 2001, we obtained $X_e^2 = 97.51$ with $df = 21$. Both statistics indicate that the two factor model is rejected. If we want to investigate the reason of the poor fit we can look at the GFfits for each pair of items and follow the rule of thumb by which a cell greater than 4 or a total greater than 36 is an indication of poor fit (Jöreskog and Moustaki, 2001). In Table 2.2 the values of the GF fits are reported.

Table 2.2 Bivariate GF fit, years 2000–2001, NLSF

	2000				2001			
StudUnc	–							
CampUnc	53.74	–			56.28	–		
BadProf	17.36	8.76	–		15.71	25.06	–	
DiscProf	19.58	19.94	13.29	–	9.04	26.82	20.07	–

We can observe that the items responsible for bad fit are StudUnc and CampUnc since the value of GF fit is greater than 36.

These results suggest that a longitudinal analysis should be performed only for the second latent variable, that we can interpret as "Professor Prejudice", since it is well measured by the items Badprof and Discprof. In particular, we want to evaluate if there is a significant change over time of this latent variable.

One fundamental assumption in latent variable models for longitudinal data is the measurement invariance of thresholds and loadings, that is the thresholds and the loadings have to be constrained to be invariant for the same item over time. We first fitted the model described in Sect. 1.4 without imposing any equality measurement constraints (Jöreskog, 2002) but the algorithm did not converge. Then we fitted the model with constrained loadings (ModA) and constrained thresholds and loadings (ModB) and in both cases the algorithm converged. However we found that the latter model has a lower BIC than the former (39495.96 for ModA versus 28183.64 for ModB). The results for ModB are reported in Table 2.3.

We fixed to 1 the loading associated to the same item in the two time points so that the latent variable is identified over time. However the loading estimate associated to DiscProf is very close to 1 and significant, indicating that the two items have the same influence on the latent variable. The variances of the random effects are

Table 2.3 Estimated thresholds and factor loadings with standard errors in brackets for the non-stationary model, NLSF

Items	$\hat{\tau}_{i(1)}$	$\hat{\tau}_{i(2)}$	$\hat{\alpha}_i$	$\hat{\sigma}_{ui}$
Badprof	4.98 (0.08)	7.00 (0.12)	1.00	0.60 (0.13)
Discprof	5.09 (0.08)	6.65 (0.13)	0.92 (0.19)	0.45 (0.06)

significant too, that implies that the random effects explain significantly the variability of the items over time.

The estimated covariance matrix of the latent variable over time is

$$\hat{\Gamma} = \begin{bmatrix} 8.04 & 7.43 \\ 7.43 & 7.86 \end{bmatrix}$$

and the estimated $\hat{\phi} = 0.92(0.03)$ shows a very strong significant correlation between the latent variables in the two time points. Moreover the variability of the latent variable decreases over time.

The results highlight that the two items Badprof and Discprof measure the latent construct with almost the same magnitude. Moreover the perception of prejudice of the students towards the professors does not change substantially over the two observed years.

2.6 Concluding remarks

Latent variable models for ordinal data have been discussed with particular attention to two aspects recently developed, the goodness-of-fit problem and the analysis of longitudinal data. As for the former, a test based on bivariate marginal distributions has been presented. It allows to overcome the sparseness problem, typical of categorical data, that invalidates the classical goodness of fit statistics.

As for the latter, model for ordinal data have been extended to longitudinal data in such a way that different kinds of variability present in the data can be modelled. At this regard the associations among items are explained by means of time dependent latent variables. A non-stationary autoregressive structure allows to evaluate their changes over time. Random effect components capture the variability of the same item over time. The potentiality of this model and the validity of the goodness-of-fit test based on residuals in presence of sparse data have been showed by means of a full application to a subset of the NLSF.

Chapter 3
Issues on item response theory modelling

Mariagiulia Matteucci, Stefania Mignani and Bernard P. Veldkamp

3.1 Introduction

Item response theory (IRT) models have been developed in order to study the individual responses to a set of items designed to measure latent abilities. IRT is a measurement theory that was first formalized in the sixties with the fundamental work of Lord and Novick (1968) and it has a predominant role in educational assessment (see van der Linden and Hambleton, 1997). An IRT model describes the relationship between the observable examinee's performance in the test, typically in the form of responses to categorical items, and the unobservable latent ability. Therefore, IRT models can be included in the more general framework of latent variable modelling (see Skrondal and Rabe-Hesketh, 2004).

IRT is used in all phases of test administration, from the test calibration to the estimation of individual abilities, in which the estimated item parameters are used to characterize the examinees. After a brief presentation of the main assumptions of IRT models, several aspects related to specific problems in the context of test administration are treated. Many advances have been introduced over the last few years, that allow both to support more complex models and to improve the estimation algorithms. In particular, issues on multidimensionality (Wang et al., 2004), incomplete design (Béguin and Glas, 2001) and the inclusion of prior information (van der Linden, 1999) are discussed, referring both to current literature and to some

---

Mariagiulia Matteucci
Department of Statistical Sciences, University of Bologna, Via Belle Arti 41, 40126 Bologna, Italy, e-mail: m.matteucci@unibo.it

Stefania Mignani
Statistics Department "Paolo Fortunati", University of Bologna, Via Belle Arti 41, 40126 Bologna, Italy, e-mail: stefania.mignani@unibo.it

Bernard P. Veldkamp
Department of Research Methodology, Measurement and Data Analysis, University of Twente, P.O. Box 217, 7500 AE Enschede, The Netherlands, e-mail: b.p.veldkamp@gw.utwente.nl

contributions of the authors. Particular attention is given to the use of the Gibbs sampler, in the Markov chain Monte Carlo (MCMC) methods, for the estimation of IRT models (Albert, 1992; Fox and Glas, 2001). Finally, an application related to one of this topics is presented in the context of educational assessment.

3.2 Basics of item response theory

Item response theory (IRT) has been developed in the field of psychometrics and has been intensely applied since the 1990s, especially in the field of educational assessment. The roots of IRT can be traced back in the thirties and forties but the theoretical work was formalized for the first time in the sixties with the fundamental work of Lord and Novick (1968). IRT has been developed for overcoming the lacks of classical test theory (CTT) (Novick, 1966), especially in terms of sensitivity to sample conditions. Nowadays, the use of IRT is widespread but researchers are advised to make a complementary use of both methodologies in order to get well-founded results.

As a measurement theory, IRT focuses on the relationship between the examinees' performance in a test and the latent ability(ies) and it is included in the framework of latent variable modelling (see Bartholomew and Knott, 1999; Skrondal and Rabe-Hesketh, 2004). In this context, IRT is used with categorical observed and continuous latent variables and it is sometimes named latent trait analysis (LTA), where the concept of trait stands for ability. An IRT model is a mathematical function used to describe the trace line(s) or conditional probability of a response given the latent variable, for an item with categorical responses (Thissen and Steinberg, 1986). Essentially, the parametric model describes the relationship between the *observable* (the individual responses to the items in the test) and the *unobservable* (the latent ability). IRT models are founded on the assumption of local independence which implies that, when the latent space has been completely specified, the examinees' responses to a set of items are statistically independent. Another ordinary assumption is unidimensionality, with reference to the presence of a single trait affecting the test performance. This condition is hardly achievable in practice, because examinees usually employ different abilities to answer a set of items. Nevertheless, what is required for unidimensionality is the existence of a single dominant component characterizing the responses. In Sect. 3.3.1 this assumption will be relaxed with the introduction of multidimensional models.

The choice of IRT model depends on the data structure. Items can have only two response categories (correct and incorrect) or more than two (nominal or ordinal). In the first case, models for binary data should be employed, while in the second one, models for polytomous data are recommended. Different models may be obtained by using different mathematical functions to model the relation between the performance and the ability, and varying the number of item parameters. Examples of application of IRT models in educational and vocational testing may be found in Matteucci and Stracqualursi (2006); Matteucci et al. (2008).

3 Issues on item response theory modelling

Consider a set of binary items and suppose the existence of a single latent ability η underlying the response process. A unidimensional IRT model for binary data expresses the probability of a correct response for each item as a function of the ability and a set of item parameters. Most common probability models make use of the logistic or the normal distribution function. When the distribution is logistic and two item parameters are considered in the model, the two-parameter logistic (2PL) model (Birnbaum, 1968) specifies the probability of a correct response for the individual i on item k, with $i = 1,...,n$ examinees and $k = 1,...,K$ items, as follows

$$\Pr(Y_{ik} = 1|\eta_i, \alpha_k, \delta_k) = \frac{\exp(\alpha_k \eta_i - \delta_k)}{1 + \exp(\alpha_k \eta_i - \delta_k)}, \quad (3.1)$$

where Y_{ik} is the binary response variable of individual i to item k, α_k and δ_k are the item parameters, and η_i is the latent ability of person i. Traditionally, the ability is denoted by the Greek letter θ in IRT, but it has been changed into η in order to keep the notation consistent within the book. Sometimes, the linear predictor is denoted as $\alpha_k(\eta_i - \beta_k)$, where α_k is the item discrimination and β_k is the item difficulty. In model (3.1), the parameter δ_k is the negative of the intercept or $\delta_k = \alpha_k \beta_k$; the parameter δ_k is called item difficulty as well.

When the normal distribution is employed, the two-parameter normal ogive (2PNO) model (Lord, 1952) is obtained as follows

$$\Pr(Y_{ik} = 1|\eta_i, \alpha_k, \delta_k) = \Phi(\alpha_k \eta_i - \delta_k) = \int_{-\infty}^{\alpha_k \eta_i - \delta_k} \frac{1}{\sqrt{2\pi}} e^{-z^2/2} dz, \quad (3.2)$$

where Φ is the standard normal cumulative distribution function.

The use of models (3.1) and (3.2) is widespread in case of dichotomous data, because they allow to study the psychometric properties of difficulty and differentiating power. In both models the probability of a correct response is expressed as a monotonically increasing function of the trait. The curve is called *item characteristic curve* (ICC) and allows a test-taker with high ability to have a high probability of endorsing the item. Haley (1952) proved that models (3.1) and (3.2) are equivalent in terms of predicting the same probability, after the introduction of a scaling constant $D = 1.702$ into the logistic model. Other popular models are the one-parameter logistic model (Rasch, 1960) and the three-parameter logistic model (Birnbaum, 1968).

The probability of observing an individual response pattern (the complete sequence of responses) can be expressed, by using the assumption of local independence, as

$$\Pr(Y_{i1},...,Y_{iK}|\eta_i, \xi) = \prod_{k=1}^{K} [\Pr(Y_{ik} = 1|\eta_i, \xi)]^{Y_{ik}} [1 - \Pr(Y_{ik} = 1|\eta_i, \xi)]^{1-Y_{ik}}, \quad (3.3)$$

where ξ is the vector of item parameters for all the K items.

Furthermore, the complete likelihood function is obtained multiplying over all the examinees, thanks to the assumption of experimental independence, such as

$$\Pr(\mathbf{Y}|\boldsymbol{\eta},\boldsymbol{\xi}) = \prod_{i=1}^{n}\prod_{k=1}^{K}[\Pr(Y_{ik}=1|\eta_i,\boldsymbol{\xi})]^{Y_{ik}}[1-\Pr(Y_{ik}=1|\eta_i,\boldsymbol{\xi})]^{1-Y_{ik}}. \quad (3.4)$$

Usually we assume that the observed data are a random sample from a population where ability is normally distributed. Clearly, from (3.1) and (3.2) we can see that both models are not univocally determined: if we multiply η_i by a constant and divide α_k by the same constant or if we add to η_i and to δ_k/α_k the same quantity, the model does not change. Constraints on item or person parameters allow to solve these two indeterminacies. Usually, in the case of unidimensionality, the model identification is conducted by fixing the mean value and the standard deviation of the ability distribution to 0 and 1, respectively.

When the item response model fits, some useful properties are achieved. First of all, item and ability estimates are said to be invariant. This property implies that the item parameter estimates are independent of the group of examinees used from the population of examinees for whom the test was designed. As a consequence, item parameters can be estimated, collected and stored in an item bank. From this item bank, new tests can be assembled for application to different populations (van der Linden, 2005). Besides, examinee ability estimates are not dependent on the particular choice of test items used from the population of items which were calibrated, i.e. which item parameters have been estimated. So, scores of examinees resulting from different tests can be compared to each other. This property is often applied in computer-based testing, e.g. in computerized adaptive testing (CAT) (Wainer et al., 1990). CAT is a form of individualized testing, where the difficulty of the items is adapted to the estimated ability level of the examinee. Another advantage is that estimates of standard errors for individual ability estimates are possible instead of a single estimate of error for all the examinees, as in the case of CTT.

3.2.1 Parameter estimation

In item response models, the probability of a correct response is a function of examinees' ability and item parameters. These two characteristics are both unknown. The only available data are the responses to a set of items given by a sample of individuals. In the estimation process, two important features should be taken into account: the nonlinearity of the response model and the impossibility of observing the latent variable η. The model estimation is analogous to performing a non-linear regression with unknown predictor values. The main focus is on the determination of the η values for each examinee and the item parameters from the item responses.

The simultaneous estimation of the ability and the item parameters can be performed according to either maximum likelihood (ML) methods or a Bayesian

3 Issues on item response theory modelling

framework. As a general rule, the estimation depends on how the probability of the observed response patterns is conceptualized. In the *stochastic subject* interpretation of probability, the observed persons are regarded as fixed. The probability represents the unpredictability of specific events, i.e. the encounter of a person with a particular item. Within this approach, the latent variables are constructed as unknown fixed parameters. In the *random sampling* interpretation of probability, the observed persons are regarded as a random sample from a population. Therefore, a specific distribution of the latent trait must be assumed to interpret the probability and the latent variables are treated as random. Three ML estimation methods are available:

- Joint maximum likelihood (JML).
- Conditional maximum likelihood (CML).
- Marginal maximum likelihood (MML).

The first two methods imply the concept of fixed latent variable while in the MML estimation the latent variables are treated as random.

The JML implements the maximum likelihood through an iterative procedure to estimate the item parameters and the abilities simultaneously. Simply, we look for the values of the parameters that jointly maximize the log-likelihood function. After the specification of the starting values, the item parameter estimates and the ability estimates are alternatively updated. This method is very simple to implement but the complexity increases as the number of observations increases. The standard limit theorems do not apply and the resulting parameter estimators are not consistent.

The CML is based on the availability of a sufficient statistic for the ability so that the likelihood function can be simplified conditioning to it. This method can be applied only for models belonging to the Rasch family, as the one-parameter logistic model (Rasch, 1960) for the binary case, where a sufficient statistic for the ability is represented by the total test score. The JML and CML methods have been applied intensively in the past but are rather limited.

Nowadays, the model estimation technique which is used mostly is MML, which is based on the marginal probability of observing a response pattern, obtained integrating out over the distribution of ability as follows

$$\Pr(\mathbf{Y}) = \int_{-\infty}^{+\infty} \Pr(\mathbf{Y}, \eta) d\eta = \int_{-\infty}^{+\infty} \Pr(\mathbf{Y}|\eta) \phi(\eta) d\eta, \tag{3.5}$$

where $\phi(\eta)$ is the prior density of the latent ability. The MML employs the EM iterative procedure (Dempster et al., 1977; Bock and Aitkin, 1981), which is based on computing first the expected values of the response patterns for each item, conditioned on the data and current parameter estimates, and subsequently on the maximization of the log-likelihood with respect of the item parameters using the expected values. Afterwards, a single ability value may be associated to each examinee by using, among the other techniques, maximum a posteriori (MAP) or expected a posteriori (EAP) methods.

All the ML estimators refer to fixed item parameters. On the other hand, the Bayesian approach regards both the latent variable and the item parameters as random. In the following, the focus will be on the implementation of the Gibbs sampler algorithm (Geman and Geman, 1984; Gelfand and Smith, 1990) in a Bayesian framework in order to estimate the 2PNO model (see Albert, 1992), which is easier to treat than the 2PL model within this approach. The Gibbs sampler is a member of the Markov chain Monte Carlo (MCMC) class of techniques and it is applied when the posterior distribution is high-parameterized and difficult to sample from. The basic idea of the algorithm is to subdivide the parameter vector (in the Bayesian notation, the parameters are random variables) in order to sequentially generate parameter values from the single conditional distributions. Due to the computational intensity that limited its use in the past, the Gibbs sampler is a relatively new estimation method in IRT but it is now widely employed thanks to modern technologies.

In order to use the Gibbs sampler for the estimation of item and person parameters in model (3.2), we need to model the presence of the dichotomous variable Y_{ik}, indicating correct or incorrect response of person i to item k, through the introduction of the underlying variables Y_{ik}^*, independent and identically distributed as $Y_{ik}^* \sim N(\alpha_k \eta_i - \delta_k; 1)$, with $i = 1, ..., n$ individuals and $k = 1, ..., K$ items. Afterwards, we should be able to simulate from the joint posterior distribution of $(\mathbf{Y}^*, \boldsymbol{\eta}, \boldsymbol{\xi})$ using the following assumptions:

- $\{Y_{ik}^*\}$ i.i.d.$\sim N(\zeta_{ik}, 1)$, with $\zeta_{ik} = \alpha_k \eta_i - \delta_k$.
- $\{Y_{ik}\}$ indicators of values of $\{Y_{ik}^*\}$.
- standard normal prior distribution on ability: $\{\eta_i\}$ i.i.d.$\sim N(0,1)$.
- prior distribution on item parameters: $\Pr(\boldsymbol{\xi}) = \prod_{k=1}^{K} I(\alpha_k > 0)$.

The last assumption insures that the discrimination parameters are positive to preserve the increasing monotonic trend of the item characteristic curve. Thus, the joint posterior distribution is given by

$$\Pr(\mathbf{Y}^*, \boldsymbol{\eta}, \boldsymbol{\xi} | \mathbf{Y}) = \Pr(\mathbf{Y}^* | \boldsymbol{\eta}, \boldsymbol{\xi}, \mathbf{Y}) \Pr(\boldsymbol{\eta}) \Pr(\boldsymbol{\xi})$$

$$\propto \prod_{i=1}^{n} \prod_{k=1}^{K} \{\phi(Y_{ik}^*; \zeta_{ik}, 1) [I(Y_{ik}^* > 0) I(Y_{ik} = 1)$$

$$+ I(Y_{ik}^* \leq 0) I(Y_{ik} = 0)]\}$$

$$\prod_{i=1}^{n} \phi(\eta_i; 0, 1) \prod_{k=1}^{K} I(\alpha_k > 0), \quad (3.6)$$

where $I(\cdot)$ is the indicator function, taking value 1 when the argument is true and 0 otherwise. Because of the intractable form of (3.6) we can resort to the Gibbs sampler using the conditional distributions of $\mathbf{Y}^*$, $\boldsymbol{\eta}$ and $\boldsymbol{\xi}$, respectively $Pr(\mathbf{Y}^* | \boldsymbol{\eta}, \boldsymbol{\xi}, \mathbf{Y})$, $Pr(\boldsymbol{\eta} | \mathbf{Y}^*, \boldsymbol{\xi}, \mathbf{Y})$ and $Pr(\boldsymbol{\xi} | \mathbf{Y}^*, \boldsymbol{\eta}, \mathbf{Y})$, which are tractable and easy to draw samples from. The conditional distribution of the independent Y_{ik}^* is normal, with expected value $\zeta_{ik} = \alpha_k \eta_i - \delta_k$ and variance equal to 1, truncated by 0 to the left if $Y_{ik} = 1$ and to the right if $Y_{ik} = 0$, as follows

$$Y_{ik}^* | \boldsymbol{\eta}, \boldsymbol{\xi}, \mathbf{Y} \sim \begin{cases} N(\zeta_{ik}, 1) \text{ with } Y_{ik}^* > 0 \text{ if } Y_{ik} = 1, \\ N(\zeta_{ik}, 1) \text{ with } Y_{ik}^* \leq 0 \text{ if } Y_{ik} = 0. \end{cases} \quad (3.7)$$

3 Issues on item response theory modelling

The conditional distribution of $\boldsymbol{\eta}$ is also normal: the person parameters $\eta_1,\dots,\eta_n$ are independent with the following conditional posterior distribution

$$\Pr(\eta_i|\mathbf{Y}^*,\boldsymbol{\xi},\mathbf{Y}) \propto \prod_{k=1}^{K} \phi(Y_{ik}^*; \zeta_{ik}, 1)\, \phi(\eta_i; 0, 1). \tag{3.8}$$

A normal regression model is assumed for observation i in the form of

$$\begin{aligned} Y_{ik}^* &= \alpha_k \eta_i - \delta_k + \upsilon_{ik} \\ Y_{ik}^* + \delta_k &= \alpha_k \eta_i + \upsilon_{ik}, \end{aligned} \tag{3.9}$$

where υ_{ik} i.i.d. $\sim N(0,1)$. The second formulation of (3.9) can be interpreted as the multiple regression of $(Y_{ik}^* + \delta_k)$ on the regressors α_k, with $k = 1,\dots,K$, considering the η_i as regression coefficients. On the assumption of standard normal prior distribution for η_i, we need to combine the likelihood and prior distribution information together in a normal model. The likelihood function of η_i follows a normal distribution with mean equal to the least square estimate of η_i, specifically $\hat{\eta}_i = \sum_{k=1}^{K} \alpha_k (Y_{ik}^* + \delta_k) / \sum_{k=1}^{K} \alpha_k^2$, and variance $v = 1/\sum_{k=1}^{K} \alpha_k^2$. Therefore, the combination of our standard normal prior distribution and the likelihood results, leads to the following normal posterior distribution for η_i

$$\eta_i|\mathbf{Y}^*,\boldsymbol{\xi},\mathbf{Y} \sim N\left(\frac{\hat{\eta}_i/v}{1/v+1}; \frac{1}{1/v+1}\right). \tag{3.10}$$

The third conditional distribution $Pr(\boldsymbol{\xi}|\mathbf{Y}^*,\boldsymbol{\eta},\mathbf{Y})$ can be computed by using the same approach applied to the fully conditional distribution of η_i. Consider the K vectors of item parameters $\boldsymbol{\xi}_1,\dots,\boldsymbol{\xi}_K$, with $\boldsymbol{\xi}_k' = [\alpha_k; \delta_k]$, independent with the following posterior distribution

$$\Pr(\boldsymbol{\xi}_k|\mathbf{Y}^*,\boldsymbol{\eta},\mathbf{Y}) \propto \prod_{i=1}^{n} \phi(Y_{ik}^*; \zeta_{ik}, 1)\, I(\alpha_k > 0). \tag{3.11}$$

The normal regression model for each item k, with $k = 1,\dots,K$, is

$$\mathbf{Y}_k^* = [\boldsymbol{\eta}\ -\mathbf{1}]\boldsymbol{\xi}_k + \mathbf{v}_k, \tag{3.12}$$

where $\boldsymbol{\eta}$ is the n-dimensional vector of individual abilities, $-\mathbf{1}$ is a n-dimensional vector with entries equal to -1 and $\mathbf{v}_k = (\upsilon_{1k},\dots,\upsilon_{nk})$ is a random sample from a standard normal distribution. The model can be interpreted as the regression of $\mathbf{Y}_k^*$ on the explanatory variables $\mathbf{U} = [\boldsymbol{\eta}\ -\mathbf{1}]$, considering the $\boldsymbol{\xi}_k$ as regression coefficients. Therefore, the likelihood function of $\boldsymbol{\xi}_k$ follows the normal distribution with mean equal to the usual least squares estimate $\hat{\boldsymbol{\xi}}_k = (\mathbf{U}'\mathbf{U})^{-1}\mathbf{U}'\mathbf{Y}_k^*$ and variance equal to $(\mathbf{U}'\mathbf{U})^{-1}$. Consequently, the posterior distribution obtained combining the likelihood function and the prior distribution on item parameters $\boldsymbol{\xi}_k$ is given by

$$\boldsymbol{\xi}_k|\mathbf{Y}^*,\boldsymbol{\eta},\mathbf{Y} \sim N(\hat{\boldsymbol{\xi}}_k; (\mathbf{U}'\mathbf{U})^{-1})\, I(\alpha_k > 0). \tag{3.13}$$

Another possible solution for computing the posterior density is to choose a prior covariance matrix for the item parameters denoted by

$$\boldsymbol{\Sigma}_0 = \begin{pmatrix} s_\alpha^2 & 0 \\ 0 & s_\delta^2 \end{pmatrix},$$

where s_α and s_δ are the prior standard deviations for α_k and δ_k. Therefore, the conditional posterior distribution of $\boldsymbol{\xi}_k$ is a multivariate normal with mean vector equal to $\left(\mathbf{U}'\mathbf{U} + \boldsymbol{\Sigma}_0^{-1}\right)^{-1} \mathbf{U}'\mathbf{Y}_k^*$ and covariance matrix equal to $\left(\mathbf{U}'\mathbf{U} + \boldsymbol{\Sigma}_0^{-1}\right)^{-1}$.

After the specification of the single conditional distributions for $\mathbf{Y}^*$, $\boldsymbol{\eta}$ and $\boldsymbol{\xi}$, it is possible to implement the Gibbs sampler to generate a sequence of drawings from these distributions in three steps:

1. Start with initial values $\boldsymbol{\xi}^{(0)}$, $\boldsymbol{\eta}^{(0)}$ and sample $\mathbf{Y}^{*(0)}$ from $\Pr(\mathbf{Y}^*|\boldsymbol{\eta},\boldsymbol{\xi},\mathbf{Y})$.
2. Use $\mathbf{Y}^{*(0)}$, $\boldsymbol{\xi}^{(0)}$ and sample $\boldsymbol{\eta}^{(1)}$ from $\Pr(\boldsymbol{\eta}|\mathbf{Y}^*,\boldsymbol{\xi},\mathbf{Y})$.
3. Use $\mathbf{Y}^{*(0)}$, $\boldsymbol{\eta}^{(1)}$ and sample $\boldsymbol{\xi}^{(1)}$ from $\Pr(\boldsymbol{\xi}|\mathbf{Y}^*,\boldsymbol{\eta},\mathbf{Y})$.

Steps 1–3 should be repeated iteratively until convergence. Generally, the MCMC sampling procedures are not sensitive to the choice of starting values; however, reasonable initial values can reduce the time of convergence. Coherently with the prior assumptions about the ηs, a possible solution is to initialize the ability parameters to their prior mean, which is equal to 0. According to Albert (1992), starting values for the item parameters α_k and δ_k, can be set to 2 and $-\Phi^{-1}\left[(\hat{p}_k)\sqrt{5}\right]$, respectively, where $\hat{p}_k = \sum_i Y_{ik}/n$ is the proportion of correct answers for each item k. However, one can decide to initialize the item parameters to suitable values, according to prior knowledge. For example, we may expect that the discrimination parameters vary between 0 and 2 and since the difficulty parameters are on the real line, a possible initialiazion is to set all αs to 1 and all δs to 0 (see Béguin and Glas, 2001). Another possibility is to use the marginal maximum likelihood (MML) parameter estimates, but the procedure requires the implementation of the EM algorithm.

3.3 Advances in IRT: some issues

For a long time, the focus of item response theory has been on the estimation of models for binary and polytomous (nominal or ordinal) data, under the assumption of unidimensionality. Within this assumption, a large variety of models with different features has been developed (Thissen and Steinberg, 1986). However, the idea of a multidimensional latent space has occurred in the IRT specific literature with the formulation of the multidimensional two-parameter normal ogive model due to Lord (1952) and Lord and Novick (1968). Despite the wide interest shown for this approach, the use of multidimensional models has been limited in practice, due both to the computational intensity of the estimation procedures and to the difficulty of making this approach easily employable in practice. Recently, general models have been proposed, that allow for the presence of more than one latent trait (Adams et al., 1997; Segall, 1996; Veldkamp, 1999; Veldkamp and van der Linden, 2002; Wang et al., 2004), and are characterized by high flexibility. Furthermore, the introduction of modern computers has facilitated the implementation of algorithms based on MCMC simulation, such as the Gibbs sampler (see Sect. 3.2.1), which, despite the computational intensity, may be applied easily to a wide range of complex models. Besides the issue of multidimensionaly, the Gibbs sampler has shown great potential in treating several issues in test administration, like missing data due to incomplete designs and the use of prior information. In this section, these issues will be discussed within the framework of Bayesian parameter estimation.

3.3.1 Multidimensionality

The detection of dimensionality is a crucial issue in IRT and in latent variable modelling. Many IRT models are based on the assumption of unidimensionality, which refers to the presence of a single latent ability. Especially when a test contains mutually exclusive subsets of items or when the underlying dimensions are not highly correlated, the use of a unidimensional model can bias the parameter estimation and the trait administration. Violation of the assumption of unidimensionality may affect the further assumption of local independence, which is valid only when the complete latent space has been specified. For this reason, many researchers tried to develop methods for the detection of dimensionality. Among others, exploratory factor analysis is used intensively in order to detect the dimensionality of a test structure. Often in educational testing, the plot of eigenvalues shows only one dominant factor, which accounts for a small percentage of variability. In the classical test theory, the reliability coefficient Alpha is used to investigate the internal consistency of the test and high value is an indicator of unidimensionality. Despite all the efforts in detecting the correct dimensionality of a test, a single and effective statistical procedure is not available yet, due to the latent nature of the phenomenon and the impossibility of comparison with observed results. Nowadays, testing is also oriented towards the evaluation of multiple competencies and the use of multidimensional models has become very important. The contemporary presence of more that one latent trait increases the complexity of the model but allows a deeply investigation of the data structure. A generalization of model (3.2) to the presence of more than one single trait is the multidimensional two-parameter normal ogive model (Lord, 1952; Lord and Novick, 1968), given by

$$Pr(Y_{ik} = 1 | \boldsymbol{\eta}_i) = \Phi\left(\sum_{q=1}^{Q} \alpha_{kq} \eta_{iq} - \delta_k\right), \qquad (3.14)$$

where $\boldsymbol{\eta}_i = (\eta_{i1}, ..., \eta_{iQ})'$ is the vector of length Q of abilities for the individual i. Model (3.14) is referred as a *compensatory* model, because a low value on one ability can be compensated by a higher value on another dimension.

A necessary condition for identification of model (3.14) is that a scale must be assigned to each latent variable. Another issue is latent variable indeterminacy, since different rotations for the discrimination parameters are possible. There are two different approaches to identify the model: the first one is to constrain the ability parameters while the other one is to act on the item parameters. The first solution consists of setting the mean and the variance-covariance matrix of the latent abilities equal to a vector of zeros and to the identity matrix, respectively. Furthermore one has to fix $\alpha_{kq} = 0$ for $k = 1, ..., Q-1$ and $q = k+1, ..., Q$. Another way to identify the model is to set some restrictions only on item parameters, that is

- Impose $\alpha_{kq} = 1$ if $k = q$ and $\alpha_{kq} = 0$ if $k \neq q$, for $k = 1, ..., Q$ and $q = 1, ..., Q$.
- Set Q item difficulties δ_k equal to 0.

Béguin and Glas (2001) proved that the two identification methods are interchangeable; however, because in multidimensional models the interpretation of the ability is more complicated, it is convenient to impose constraints on item parameters, rather than on person parameters. Béguin and Glas (2001) extended the implementation of the Gibbs sampler by Albert (1992) to multidimensional IRT models. In particular, they implemented the algorithm for the three-parameter normal ogive (3PNO) model, which includes a guessing parameter. Following Albert (1992) and Béguin and Glas (2001), the Gibbs sampler has been implemented for the multidimensional 2PNO (Matteucci, 2007a, b). The focus has been on the 2PNO model respect to the 3PNO because, even if the presence of a guessing factor is intrinsic to multiple-choice items, the 3PNO model presents several estimation problems and it is not so flexible to be included in the generalized linear latent variable models (GLLVM).

Assuming the existence of an underlying continuous response variable Y_{ik}^*, the posterior distribution of interest is $\Pr(\mathbf{Y}^*, \boldsymbol{\eta}, \boldsymbol{\xi}, \boldsymbol{\mu}, \boldsymbol{\Sigma}_\eta | \mathbf{Y})$, where $\mathbf{Y}^*$ is the $n \times K$ matrix of the underlying variables, $\boldsymbol{\eta}$ is the $n \times Q$ matrix of abilities, $\boldsymbol{\xi}$ is the $K \times (Q+1)$ matrix of item parameters, $\boldsymbol{\mu}$ and $\boldsymbol{\Sigma}_\eta$ are the mean and variance-covariance matrix of the latent abilities, respectively. A multivariate normal distribution is assumed for the abilities, i.e. $\boldsymbol{\eta}_1, ..., \boldsymbol{\eta}_n$ are independent and $\boldsymbol{\eta}_i \sim N(\boldsymbol{\mu}, \boldsymbol{\Sigma}_\eta)$. Furthermore, a multivariate normal and an inverse-Wishart prior distributions for $\boldsymbol{\mu}$ and $\boldsymbol{\Sigma}_\eta$ are considered, respectively: $\boldsymbol{\mu} | \boldsymbol{\Sigma}_\eta \sim N\left(\boldsymbol{\mu}_0, \frac{\boldsymbol{\Sigma}_\eta}{\kappa_0}\right)$, where $\boldsymbol{\mu}_0$ is the prior mean vector and κ_0 is the number of prior measurements on the $\boldsymbol{\Sigma}_\eta$ scale, and $\boldsymbol{\Sigma}_\eta \sim \text{Inv-Wishart}_{\nu_0}\left(\boldsymbol{\Lambda}_0^{-1}\right)$, with ν_0 degrees of freedom and a symmetric, positive definite scale matrix $\boldsymbol{\Lambda}_0$. Finally, normal distributions are assumed for the item parameters: $\alpha \sim N\left(\mu_\alpha, \sigma_\alpha^2\right)$ and $\delta \sim N\left(\mu_\delta, \sigma_\delta^2\right)$.

As for the unidimensional model, it is possible to express the joint posterior distribution of $\mathbf{Y}^*, \boldsymbol{\eta}, \boldsymbol{\xi}, \boldsymbol{\mu}$ and $\boldsymbol{\Sigma}_\eta$ as follows

3 Issues on item response theory modelling

$$\Pr(\mathbf{Y}^*, \boldsymbol{\eta}, \boldsymbol{\xi}, \boldsymbol{\mu}, \boldsymbol{\Sigma}_\eta | \mathbf{Y}) = \Pr(\mathbf{Y}^* | \mathbf{Y}, \boldsymbol{\xi}, \boldsymbol{\eta}) \Pr(\boldsymbol{\eta} | \boldsymbol{\mu}, \boldsymbol{\Sigma}_\eta) \Pr(\boldsymbol{\mu} | \boldsymbol{\Sigma}_\eta) \Pr(\boldsymbol{\Sigma}_\eta) \Pr(\boldsymbol{\xi})$$

$$\propto \prod_{i=1}^{n} \prod_{k=1}^{K} \{\phi(Y_{ik}^*; \zeta_{ik}, 1) \left[I(Y_{ik}^* > 0) I(Y_{ik} = 1) + I(Y_{ik}^* \leq 0) I(Y_{ik} = 0) \right] \}$$

$$\prod_{i=1}^{n} \phi(\boldsymbol{\eta}_i; \boldsymbol{\mu}, \boldsymbol{\Sigma}_\eta) \Pr(\boldsymbol{\mu} | \boldsymbol{\Sigma}_\eta) \Pr(\boldsymbol{\Sigma}_\eta) Pr(\boldsymbol{\xi}). \tag{3.15}$$

Assuming initial appropriate estimates for $\boldsymbol{\eta}$ and $\boldsymbol{\xi}$, the Gibbs sampler works with the conditional densities $\Pr(\boldsymbol{\mu}, \boldsymbol{\Sigma}_\eta | \boldsymbol{\eta})$, $\Pr(\mathbf{Y}^* | \boldsymbol{\eta}, \boldsymbol{\xi}, \mathbf{Y})$, $\Pr(\boldsymbol{\eta} | \mathbf{Y}^*, \boldsymbol{\xi}, \mathbf{Y}, \boldsymbol{\mu}, \boldsymbol{\Sigma}_\eta)$ and $\Pr(\boldsymbol{\xi} | \mathbf{Y}^*, \boldsymbol{\eta}, \mathbf{Y})$.

Combining the information of the prior distributions and the likelihood function, the first conditional distribution $Pr(\boldsymbol{\mu}, \boldsymbol{\Sigma}_\eta | \boldsymbol{\eta})$ turns out to be a normal-inverse-Wishart with parameters $\boldsymbol{\mu}_n$, $\boldsymbol{\Sigma}_\eta / \kappa_n$, v_n and $\boldsymbol{\Lambda}_n$, where:

$$\boldsymbol{\mu}_n = \frac{\kappa_0}{\kappa_0 + n} \boldsymbol{\mu}_0 + \frac{n}{\kappa_0 + n} \bar{\boldsymbol{\eta}}$$

$$\kappa_n = \kappa_0 + n$$

$$v_n = v_0 + n$$

$$\boldsymbol{\Lambda}_n = \boldsymbol{\Lambda}_0 + \mathbf{S} + \frac{\kappa_0 n}{\kappa_0 + n} (\bar{\boldsymbol{\eta}} - \boldsymbol{\mu}_0)(\bar{\boldsymbol{\eta}} - \boldsymbol{\mu}_0)'.$$

The matrix $\mathbf{S}$ is the $Q \times Q$ sum of squares matrix relative to the sample mean $\bar{\boldsymbol{\eta}}$.

The second conditional distribution of interest, $\Pr(\mathbf{Y}^* | \boldsymbol{\eta}, \boldsymbol{\xi}, \mathbf{Y})$, is a truncated normal, particularly

$$Y_{ik}^* | \boldsymbol{\eta}, \boldsymbol{\xi}, \mathbf{Y} \sim \begin{cases} N(\zeta_{ik}, 1) \text{ with } Y_{ik}^* > 0 \text{ if } Y_{ik} = 1, \\ N(\zeta_{ik}, 1) \text{ with } Y_{ik}^* \leq 0 \text{ if } Y_{ik} = 0, \end{cases} \tag{3.16}$$

where $\zeta_{ik} = \sum_{q=1}^{Q} \alpha_{kq} \eta_{iq} - \delta_k$.

Due to a transformation on the vector of ability parameters $\boldsymbol{\eta}_i = \boldsymbol{\mu} + \mathbf{L}\boldsymbol{\eta}_i^o$, with $\boldsymbol{\Sigma}_\eta = \mathbf{L}\mathbf{L}'$, and the normal regression interpretation of the model, the conditional distribution of $\Pr(\boldsymbol{\eta} | \mathbf{Y}^*, \boldsymbol{\xi}, \mathbf{Y}, \boldsymbol{\mu}, \boldsymbol{\Sigma}_\eta)$ is obtained as follows

$$\boldsymbol{\eta}_i^o | \mathbf{Y}^*, \boldsymbol{\xi}, \mathbf{Y} \sim N\left((\mathbf{I} + \boldsymbol{\Sigma}^{-1})^{-1} \boldsymbol{\Sigma}^{-1} \hat{\boldsymbol{\eta}}_i^o; (\mathbf{I} + \boldsymbol{\Sigma}^{-1})^{-1}\right), \tag{3.17}$$

where $\boldsymbol{\Sigma} = (\mathbf{B}'\mathbf{B})^{-1}$, with $\mathbf{B} = \mathbf{A}\mathbf{L}$, $\mathbf{A}$ is the $K \times Q$ discrimination matrix and $\hat{\boldsymbol{\eta}}_i^o = (\mathbf{B}'\mathbf{B})^{-1}\mathbf{B}'(\mathbf{Y}_i^* + \boldsymbol{\delta} - \mathbf{A}\boldsymbol{\mu})$.

Finally, following the normal assumption of item parameters, it is possible to express the prior distribution of the item parameters for the item k, with $k = 1, ..., K$, as a multivariate normal distribution. Particularly, the vector of item parameters $\boldsymbol{\xi}_k = (\alpha_{k1}, \alpha_{k2}, ..., \alpha_{kQ}, \delta_k)'$ has a multivariate normal distribution with a mean vector equal to $\boldsymbol{\mu}_{\xi_0} = (\mu_{\alpha 1}, ..., \mu_{\alpha Q}, \mu_\delta)'$ and variance $\boldsymbol{\Sigma}_{\xi_0} = \text{diag}(\sigma_{\alpha 1}, ..., \sigma_{\alpha Q}, \sigma_\delta)$. Therefore, the conditional distribution of the item parameters can be expressed as

$$\boldsymbol{\xi}_k | \mathbf{Y}^*, \boldsymbol{\eta}, \mathbf{Y} \sim N\left(\boldsymbol{\mu}_{\xi_k}; \left(\boldsymbol{\Sigma}_{\xi_0}^{-1} + \mathbf{U}'\mathbf{U}\right)^{-1}\right), \tag{3.18}$$

where $\boldsymbol{\mu}_{\boldsymbol{\xi}_k} = \left(\boldsymbol{\Sigma}_{\xi_0}^{-1} + \mathbf{U}'\mathbf{U}\right)^{-1} \left(\boldsymbol{\Sigma}_{\xi_0}^{-1} \boldsymbol{\mu}_{\xi_0} + \mathbf{U}'\mathbf{Y}^*_k\right)$ and $\mathbf{U}$ is a $n \times (Q+1)$ matrix with the η_{iq} and a column with elements equal to -1.

The Gibbs sampler can be implemented to generate a sequence of drawings from the conditional distributions in the following four steps:

1. Starting with an initial value $\boldsymbol{\eta}^{(0)}$, sample $\boldsymbol{\Sigma}_\eta^{(0)}$ from $\boldsymbol{\Sigma}_\eta \sim \text{Inv-Wishart}_{v_n}\left(\boldsymbol{\Lambda}_n^{-1}\right)$ and then sample $\boldsymbol{\mu}^{(0)}$ from $\boldsymbol{\mu}|\boldsymbol{\Sigma}_\eta, \boldsymbol{\eta} \sim N\left(\boldsymbol{\mu}_n, \frac{\boldsymbol{\Sigma}_\eta}{\kappa_n}\right)$.
2. Start with initial values of $\boldsymbol{\xi}^{(0)}$ and $\boldsymbol{\eta}^{(0)}$ and sample $\mathbf{Y}^{*(0)}$ from $\Pr(\mathbf{Y}^*|\boldsymbol{\eta}, \boldsymbol{\xi}, \mathbf{Y})$.
3. Use $\mathbf{Y}^{*(0)}$, $\boldsymbol{\xi}^{(0)}$, $\boldsymbol{\Sigma}_\eta^{(0)}$ and $\boldsymbol{\mu}^{(0)}$ to sample $\boldsymbol{\eta}^{(1)}$ from $\Pr(\boldsymbol{\eta}|\mathbf{Y}^*, \boldsymbol{\xi}, \mathbf{Y}, \boldsymbol{\mu}, \boldsymbol{\Sigma}_\eta)$.
4. Use $\mathbf{Y}^{*(0)}$ and $\boldsymbol{\eta}^{(1)}$ to sample $\boldsymbol{\xi}^{(1)}$ from $\Pr(\boldsymbol{\xi}|\mathbf{Y}^*, \boldsymbol{\eta}, \mathbf{Y})$.

Steps 1–4 should be repeated with the updated values, iteratively.

3.3.2 Incomplete design

So far we have supposed that all the data are available, that is the $\mathbf{Y}$ data matrix is complete and consists of correct and wrong responses coded by 1 and 0, respectively. In practice, item responses may contain missing data due to respondents or test administrators. In the first case, the missing response may be considered as a wrong response in the educational field. In the second case the test design is incomplete because not all the items are submitted to the totality of the candidates, and the missing due to not-presented items should not be included in the estimation procedure.

The incomplete design can be implemented in the Gibbs sampler, imposing the algorithm to skip the missing data, as suggested in Béguin and Glas (2001). Next to the data matrix $\mathbf{Y}$, which contains correct, incorrect and missing responses corresponding to n examinees and K items, we can create a new matrix $\mathbf{D}$ as indicator of the incomplete design. Particularly, we have

$$d_{ik} = \begin{cases} 1 & \text{if the item } k \text{ is administered to the respondent } i, \\ 0 & \text{otherwise.} \end{cases} \quad (3.19)$$

In order to estimate the unidimensional model (3.2), steps 1–3 of Sect. 3.2.1 are replaced as follows

1. Start with initial values $\boldsymbol{\xi}^{(0)}$, $\boldsymbol{\eta}^{(0)}$ and sample $\mathbf{Y}^{*(0)}$ from $\Pr(\mathbf{Y}^*|\boldsymbol{\eta}, \boldsymbol{\xi}, \mathbf{Y})$ only for the elements equal to 1 of the $\mathbf{D}$ matrix.
2. Use $\mathbf{Y}^{*(0)}$, $\boldsymbol{\xi}^{(0)}$ and sample $\boldsymbol{\eta}^{(1)}$ from $\Pr(\boldsymbol{\eta}|\mathbf{Y}^*, \boldsymbol{\xi}, \mathbf{Y})$ conditionally on $\mathbf{D}$.
3. Use $\mathbf{Y}^{*(0)}$, $\boldsymbol{\eta}^{(1)}$ and sample $\boldsymbol{\xi}^{(1)}$ from $\Pr(\boldsymbol{\xi}|\mathbf{Y}^*, \boldsymbol{\eta}, \mathbf{Y})$ conditionally on $\mathbf{D}$.

Analogously, for the multidimensional model (3.14) the Gibbs sampler works sampling from the single conditional distributions conditionally to the $\mathbf{D}$ matrix, and steps 1–4 of Sect. 3.3.1 are replaced by the followings

3 Issues on item response theory modelling

1. Starting with an initial value $\boldsymbol{\eta}^{(0)}$, sample $\boldsymbol{\Sigma}_\eta^{(0)}$ from $\boldsymbol{\Sigma}_\eta \sim \text{Inv-Wishart}_{v_n}\left(\boldsymbol{\Lambda}_n^{-1}\right)$ and then sample $\boldsymbol{\mu}^{(0)}$ from $\boldsymbol{\mu}|\boldsymbol{\Sigma}_\eta, \boldsymbol{\eta} \sim N\left(\boldsymbol{\mu}_n, \frac{\boldsymbol{\Sigma}_\eta}{\kappa_n}\right)$.
2. Start with initial values of $\boldsymbol{\xi}^{(0)}$ and $\boldsymbol{\eta}^{(0)}$ and sample $\boldsymbol{Y}^{*(0)}$ from $\Pr(\boldsymbol{Y}^*|\boldsymbol{\eta}, \boldsymbol{\xi}, \boldsymbol{Y})$, only for the elements corresponding to $d_{ik} = 1$.
3. Use $\boldsymbol{Y}^{*(0)}$, $\boldsymbol{\xi}^{(0)}$, $\boldsymbol{\Sigma}_\eta^{(0)}$ and $\boldsymbol{\mu}^{(0)}$ to sample $\boldsymbol{\eta}^{(1)}$ from $\Pr(\boldsymbol{\eta}|\boldsymbol{Y}^*, \boldsymbol{\xi}, \boldsymbol{Y}, \boldsymbol{\mu}, \boldsymbol{\Sigma}_\eta)$, conditionally to the $\boldsymbol{D}$ matrix.
4. Use $\boldsymbol{Y}^{*(0)}$ and $\boldsymbol{\eta}^{(1)}$ to sample $\boldsymbol{\xi}^{(1)}$ from $\Pr(\boldsymbol{\xi}|\boldsymbol{Y}^*, \boldsymbol{\eta}, \boldsymbol{Y})$, conditionally to the $\boldsymbol{D}$ matrix.

The implementation and the application of the incomplete design are also discussed in Matteucci (2007a).

3.3.3 Inclusion of prior information

In testing occasions, collateral information on the candidates may be available. This may include socio-demographic variables or information about the performances obtained during other testing phases. The inclusion of background variables which are strongly related to the latent ability in the item response model may be investigated in order to improve the process of ability estimation. The introduction of prior information in the IRT model can be performed by considering a set of X_p background variables, with $p = 1, ..., P$, as follows

$$\eta_i = \beta_0 + \beta_1 X_{i1} + ... + \beta_P X_{iP} + \varepsilon_i, \quad (3.20)$$

where the error terms ε_i are i.i.d. $\sim N(0, \sigma^2)$.

Therefore, a linear relation between the background and the latent variables is considered. The conditional distribution of η_i, given the X_ps, is normal

$$\eta_i | X_{i1}, ..., X_{iP} \sim N(\beta_0 + \beta_1 X_{i1} + ... + \beta_P X_{iP}; \sigma^2). \quad (3.21)$$

The direct estimation of the βs and σ^2 may be conducted substituting (3.20) into the IRT model. In the literature, this method has been applied through MML via EM algorithm for the Rasch model (Zwinderman, 1991) and for the 2PL model (van der Linden, 1999). On the other hand, the inclusion of prior information is possible within the MCMC methods. The Gibbs sampler has been implemented to estimate a general multilevel IRT model (Fox and Glas, 2001) and to model hierarchically the measurement model and prior information in the form of response times (Fox et al., 2007).

Starting from the estimation of model (3.2), the Gibbs sampler has been extended to the inclusion of prior information on η (Matteucci and Veldkamp, 2008). As usual, from a fully Bayesian perspective, the joint posterior distribution of interest is

$$\Pr(\mathbf{Y}^*, \boldsymbol{\eta}, \boldsymbol{\xi}, \boldsymbol{\beta}, \sigma^2 | \mathbf{Y}, \mathbf{X}) = \Pr(\mathbf{Y}^* | \boldsymbol{\eta}, \boldsymbol{\xi}, \mathbf{Y}) \Pr(\boldsymbol{\eta} | \boldsymbol{\beta}, \sigma^2, \mathbf{X}) \Pr(\boldsymbol{\xi}) \Pr(\boldsymbol{\beta}) \Pr(\sigma^2). \tag{3.22}$$

Because the (3.22) has an intractable form, the Gibbs sampler algorithm can be applied in order to iteratively sample from the conditional distribution of each variable respect to all the others.

The first conditional distribution of interest is $\mathbf{Y}^* | \boldsymbol{\eta}, \boldsymbol{\xi}, \mathbf{Y}$, which, analogously to (3.7) and (3.16), is

$$Y_{ik}^* | \boldsymbol{\eta}, \boldsymbol{\xi} \sim \begin{cases} N(\zeta_{ik}, 1) \text{ with } Y_{ik}^* > 0 \text{ if } Y_{ik} = 1, \\ N(\zeta_{ik}, 1) \text{ with } Y_{ik}^* \leq 0 \text{ if } Y_{ik} = 0, \end{cases} \tag{3.23}$$

with $\zeta_{ik} = \alpha_k \eta_i - \delta_k$.

The conditional distribution of the latent ability $\boldsymbol{\eta} | \mathbf{Y}^*, \boldsymbol{\xi}, \boldsymbol{\beta}, \sigma^2$ is expressed combining the likelihood function of η_i and the prior distribution (3.21), as follows

$$\eta_i | \mathbf{Y}^*, \boldsymbol{\xi}, \boldsymbol{\beta}, \sigma^2 \sim N\left(\frac{\hat{\eta}_i/v + \mathbf{X}_i \boldsymbol{\beta}/\sigma^2}{1/v + 1/\sigma^2}; \frac{1}{1/v + 1/\sigma^2}\right), \tag{3.24}$$

where $\hat{\eta}_i = (\alpha_k' \alpha_k)^{-1} \alpha_k' (Y_{ik}^* + \delta_k)$ and $v = (\alpha_k' \alpha_k)^{-1}$.

The conditional distribution of the item parameters $\boldsymbol{\xi} | \boldsymbol{\eta}, \mathbf{Y}^*$, assuming a prior covariance matrix $\boldsymbol{\Sigma}_0$, is a multivariate normal

$$\boldsymbol{\xi}_k | \boldsymbol{\eta}, \mathbf{Y}^* \sim N\left(\left(\mathbf{W}'\mathbf{W} + \boldsymbol{\Sigma}_0^{-1}\right)^{-1} \mathbf{W}' \mathbf{Y}_k^*; \left(\mathbf{W}'\mathbf{W} + \boldsymbol{\Sigma}_0^{-1}\right)^{-1}\right), \tag{3.25}$$

where $\mathbf{W} = [\boldsymbol{\eta} \ -\mathbf{1}]$.

The remaining conditional distributions are referred to the regression parameters, specifically $\boldsymbol{\beta} | \boldsymbol{\eta}, \sigma^2$ and $\sigma^2 | \boldsymbol{\eta}, \boldsymbol{\beta}$. The former follows a multivariate normal distribution

$$\boldsymbol{\beta} | \boldsymbol{\eta}, \sigma^2 \sim N(\hat{\boldsymbol{\beta}}; \sigma^2 (\mathbf{X}'\mathbf{X})^{-1}), \tag{3.26}$$

with $\hat{\boldsymbol{\beta}} = (\mathbf{X}'\mathbf{X})^{-1} \mathbf{X}' \boldsymbol{\eta}$.

The latter follows an inverse-Chi-square distribution, specifically

$$\sigma^2 | \boldsymbol{\eta}, \boldsymbol{\beta} \sim Inv - \chi^2(n; S^2), \tag{3.27}$$

where $S^2 = \frac{1}{n}(\boldsymbol{\eta} - \mathbf{X}\boldsymbol{\beta})'(\boldsymbol{\eta} - \mathbf{X}\boldsymbol{\beta})$.

All the conditional distributions are dependent on the data $\mathbf{Y}$ and $\mathbf{X}$, which have been suppressed for notational convenience. Starting from a set of initial values, the Gibbs sampler proceeds with the iterative sampling from the single conditional distributions until convergence. The identification of the model is conducted by fixing the distribution of η so that its mean and standard deviation are 0 and 1, respectively.

3.4 Case study: prior information in educational assessment

Among the different issues described in Sect. 3.3, we have decided to focus on the use of prior information (see Sect. 3.3.3) in a case study. The aim of the application is to show how collateral information about individuals can be used in order to estimate both the measurement model and the relationship between ability and covariates in a single step. For the purpose of the analysis, we consider data from the 2008 final exam in the Italian lower secondary school. In fact, a written standardized test was submitted by the National Evaluation Institute for the School System (INVALSI) to all Italian students in the third class. The test consists of two different subscales (Mathematics and Italian) and contains multiple-choice, close constructed-response, and open constructed-response items. The Mathematics section consists of 22 items while the Italian test contains 25 items dealing with reading comprehension and grammar. Besides the responses of the candidates on all the items, also several background variables were recorded. A random stratified by geographic area, three-stage sample of 4,865 students is considered in the analysis.

The idea is to use the performance of the examinees in the Italian subscale as background information for the ability in the Mathematics subscale. Because the responses were dichotomized, maximum likelihood estimates of the ability have been computed for each subscale according to model (3.1). Besides the ability estimates in the Italian section, collateral information about gender, scholastic career and geographic origin are therefore included in the following multiple regression model

$$\eta_i = \beta_0 + \beta_1 X_{i1} + \beta_2 X_{i2} + \beta_3 X_{i3} + \beta_4 X_{i4} + \varepsilon_i, \tag{3.28}$$

where X_1 are the ML ability estimates concerning the Italian test, X_2 is the gender variable (coded 1 for females and 0 for males), X_3 is the scholastic career (coded 1 for students repeating the scholastic year at least once, and 0 otherwise) and X_4 is the indicator variable for the geographic origin (1 for foreign students and 0 for Italians).

The Gibbs sampler algorithm described in Sect. 3.3.3 has been run for 40,000 iterations, with a burn-in of 1,000 iterations. Table 3.1 contains the item parameter estimates of model (3.2) for the Mathematics test.

The parameter estimates are comparable to the results obtained by using standard IRT software. As Table 3.1 shows, the items are rather discriminating and are related to different levels of difficulties, which leads to an accurate test for the student evaluation. The chain length has been considered sufficient in order to obtain a Monte Carlo standard error which is less than 1% of the corresponding standard deviation for each parameter. The estimates of the regression parameters are reported in Table 3.2.

The regression coefficient estimates highlight a positive and large effect of the student performance in the Italian test on the Mathematics ability ($\beta_1 = 0.788$), given the standard normal scale of the trait, while moderate negative effects are reported for females ($\beta_2 = -0.305$) and students repeating the year ($\beta_3 = -0.210$). In particular, it seems that males perform slightly better than females in the test,

Table 3.1 Item parameter estimates[a]

Item	$\hat{\alpha}_k$	$\hat{\delta}_k$	Item	$\hat{\alpha}_k$	$\hat{\delta}_k$
01	0.433(0.025)	−0.842 (0.022)	12	0.487(0.025)	−0.698 (0.021)
02	0.506(0.024)	−0.467 (0.020)	13	0.559(0.025)	−0.507 (0.022)
03	0.439(0.022)	−0.141 (0.019)	14	0.432(0.022)	−0.021 (0.019)
04	0.424(0.024)	−0.632 (0.020)	15	0.523(0.023)	0.192 (0.019)
05	0.405(0.026)	0.984 (0.023)	16	0.422(0.028)	−1.112 (0.025)
06	0.902(0.033)	−0.773 (0.025)	17	0.776(0.031)	−0.781 (0.024)
07	0.828(0.032)	−0.722 (0.024)	18	0.566(0.029)	−1.069 (0.026)
08	0.645(0.026)	0.338 (0.020)	19	0.761(0.028)	−0.423 (0.021)
09	0.605(0.025)	−0.270 (0.020)	20	0.875(0.038)	1.253 (0.033)
10	0.866(0.033)	0.776 (0.025)	21	0.386(0.024)	−0.687 (0.021)
11	0.586(0.027)	0.696 (0.022)	22	0.777(0.029)	0.304 (0.020)

[a] standard deviations in brackets

Table 3.2 Regression parameters [a]

	Estimates
β_0	0.168 (0.017)
β_1	0.788 (0.014)
β_2	−0.305 (0.025)
β_3	−0.210 (0.049)
β_4	0.063 (0.057)
σ^2	0.500 (0.013)

[a] standard deviations in brackets

confirming the results obtained in educational literature. Finally, the effect of being a foreign student is estimated to be almost null ($\beta_4 = 0.063$). This seems to be a controversial result, because descriptive analysis showed that foreign students had average performances significantly lower than Italian students. This outcome is also confirmed by estimating a simple regression model with the geographic origin as the only covariate. However, when the performance in the Italian test is introduced in the model, the effect of the individual origin is absorbed by the ability covariate.

3.5 Concluding remarks

The chapter discussed the use of item response theory for the analysis of individual responses to a set of items, focusing on specific issues in the context of test administration, as multidimensionality and incomplete design, all implemented by using the Gibbs sampler. Despite the well-known properties of the MCMC methods, the estimation algorithms have been applied only recently in IRT due to their computational intensity but have shown great capability of estimation in special situations. A particular attention has been given to the inclusion of prior information in the test administration, due to its relevance for future research. In fact, the use of collateral information about the examinees may be especially useful in computerized adaptive testing (CAT), when the items are iteratively adapted to the test-taker's ability, updated after the administration of each single item. The use of prior information should be investigated in order to improve the measurement precision and to reduce the test length.

Acknowledgements The authors would like to thank the National Evaluation Institute for the School System (INVALSI) for the availability of the data used in the case study, and especially Roberto Ricci for his useful support in the data management and interpretation.

Chapter 4
Nonlinearity in the analysis of longitudinal data

Estela Bee Dagum, Silvia Bianconcini and Paola Monari

4.1 Introduction

In recent years, the use of longitudinal designs has increased appreciably and the study of change has become an essential component of research in the behavioural sciences. The availability of "micropanels", that consists of large cross-sections of individuals observed for short time periods, provides informations for answering questions about *(i)* how each individual performs over time, and *(ii)* what are the predict differences among individuals in their change. These questions form the core of every study about achievement growth, and we need suitable models to investigate the dependence structure of these longitudinal data.

Two important features have to be taken into account: *(i)* the clustering of responses within units, and *(ii)* the chronological ordering of the responses. Both *(i)* and *(ii)* imply dependence among observations on the same unit. Including random effects into statistical models is a common way of distinguishing between-subject and within-subject source of variability in view of reflecting unobserved heterogeneity in the individual behaviour (Laird and Ware, 1982). In particular, the random effects can be incorporated into Structural Equation Models (SEM) by considering them as latent variables (see e.g. Bollen and Curran (2006)). Baker (1954) was the first to suggest the use of latent variable models to study panel data, whereas Tucker (1958) gave a more technical expression of this idea for exploratory factor analysis. Meredith and Tisak (1990) took this to the confirmatory factor analysis and

---

Estela Bee Dagum
Department of Statistical Sciences, University of Bologna, via Belle Arti 41, 40126 Bologna, Italy,
e-mail: estela.beedagum@unibo.it

Silvia Bianconcini
Department of Statistical Sciences, University of Bologna, via Belle Arti 41, 40126 Bologna, Italy,
e-mail: silvia.bianconcini@unibo.it

Paola Monari
Department of Statistical Sciences, University of Bologna, via Belle Arti 41, 40126 Bologna, Italy,
e-mail: paola.monari@unibo.it

demonstrated that trajectory modelling fits naturally into this framework. We refer to these models as Latent Curve Models (LCMs). The basic idea is that individuals differ in their growth over time, and they are likely to have different temporal behaviours as a function of differences in particular characteristics, such as gender, scholar background etc. The model allows both the level of the response and the effects of covariates to vary randomly across units. In this context, an issue that researchers have to address is the nonlinearity of the functional form and particularly in the parameters. The individual growth is generally assumed to be linear, but many behavioural processes exhibit differential rates of change. Since LCM is a confirmatory factor model, it cannot estimate complex nonlinear functions directly. Hence, several methods have been developed in view of treating nonlinear dynamics via models which are linear in the parameters.

This chapter considers nonlinear latent curve models for the study of longitudinal developmental data. Section 4.2 describes linear LCM, in terms of model specification and estimation. In Sect. 4.3, we review several methods for the estimation of nonlinear LCMs. Finally, Sect. 4.4 illustrates an application on nonlinear academic performance data based on a cohort of students enrolled at the University of Bologna.

4.2 Latent curve models

In this section, we review the linear LCM to illustrate its main features. Many important texts describe a large number of extensions, and we refer the reader to Singer and Willett (2004), Skrondal and Rabe-Hesketh (2004), and Bollen and Curran, 2006.

The latent curve model posits the existence of continuous underlying trajectories which are not directly observable, but only indirectly using repeated measures. It actually consists of the following three parts:

1. The linear function that describes the change over time in the repeated measures. It is defined for individual subjects and employs subject-specific random coefficients β_{i0} and β_{i1} as follows

$$y_{it} = \beta_{i0} + \beta_{i1}(t-1) + \varepsilon_{it}, \quad t = 1, 2, ..., T_i \quad i = 1, 2, ..., n, \quad (4.1)$$

where y_{it} is the value of the response variable y for the individual i at time point t, and β_{i0} and β_{i1} are assumed to be uncorrelated among individuals.

2. The covariance structure of the residuals, also defined as within-subject structure. In Eq. (4.1), the disturbances ε_{it} are assumed to be normally distributed with zero means and heteroscedastic variances. They are also uncorrelated over time ($cov(\varepsilon_{i,t}, \varepsilon_{i,t+s}) = 0$ for $s \neq 0$), over individuals ($cov(\varepsilon_{i,t}, \varepsilon_{k,t+s}) = 0$ for $i \neq k$ and $\forall s$), and with the random coefficients ($cov(\beta_{ij}, \varepsilon_{it}) = 0, \forall j, i$).

3. The set of models for the regression coefficients β_{i0} and β_{i1} that summarize the differences between individuals. Key modelling results are estimates of the overall means, that are measures of central tendencies in the trajectories, and of the variation across individuals of the random coefficients. That is,

4 Nonlinearity in the analysis of longitudinal data

$$\beta_{ij} = \mu_{\beta_j} + \zeta_{\beta_{ij}} \quad j=0,1 \quad i=1,...,n, \qquad (4.2)$$

where the disturbances $\zeta_{\beta_{ij}}$'s are assumed to be correlated and normally distributed with zero means and variances $\psi_{\beta_j}^2$.

By incorporating covariates in Eq. (4.2), we can test for potential influences on the trajectory parameters. That is,

$$\beta_{ij} = \mu_{\beta_j} + \gamma'_{\beta_j} \mathbf{w}_i + \zeta_{\beta_{ij}}, \quad j=0,1 \quad i=1,2,...,n \qquad (4.3)$$

where γ's are the regression coefficients of the time-invariant covariates $\mathbf{w}_i$, and μ_{β_j} are mean coefficients when the covariates are zero.

Although many questions about the trajectories are possible, three are the main ones: *(i)* the characteristics of the mean trajectory of the entire group, determined by μ_{β_0} and μ_{β_1}, *(ii)* the evaluation of individual differences in trajectories picked up by the variances and covariances introduced to estimate the sampling fluctuations around the mean trajectory, and *(iii)* the potential incorporation of predictors to better understand the variability observed in the individual trajectories.

Random coefficients can be treated within the Structural Equation Modelling (SEM) perspective, where the case-specific parameters β_{i0} and β_{i1} are viewed as latent variables. They are commonly known as Latent Curve Models (LCMs).

Let the repeated measures y_{it} be stacked into the vector $\mathbf{y}$ and the latent variables β_{i0} and β_{i1} into $\boldsymbol{\eta}$, the model can be reexpressed as follows

$$\mathbf{y} = \boldsymbol{\Lambda}\boldsymbol{\eta} + \boldsymbol{\varepsilon}, \qquad (4.4)$$
$$\boldsymbol{\eta} = \boldsymbol{\tau} + \mathbf{B}\boldsymbol{\eta} + \boldsymbol{\Gamma}\mathbf{w} + \boldsymbol{\zeta}. \qquad (4.5)$$

In Eq. (4.4), $\boldsymbol{\Lambda}$ is a matrix of factor loadings, and $\boldsymbol{\varepsilon}$ is the vector of time varying errors, assumed to have zero mean and covariance matrix $\boldsymbol{\Theta}_\varepsilon$. In Eq. (4.5), $\mathbf{B}$ is a null matrix, $\boldsymbol{\tau}$ contains μ_{β_0} and μ_{β_1}, whereas $\boldsymbol{\Gamma}$ refers to the regression coefficients related to the covariates $\mathbf{w}$. The latent residual vector $\boldsymbol{\zeta}$ is assumed to be normally distributed with zero mean and covariance matrix $\boldsymbol{\Psi}$.

Differently from the classical structural equation modelling approach where the loadings are estimated, the linear LCM fixes them to specific a priori values. Moreover in SEM, the means of the factors and observed variables are usually omitted. In contrast, the LCM explicitly models both of the mean and covariance structures among the observed measures. However, a restrictive structure is imposed on these means. The intercepts of the repeated measures are set to zero, and the means for the latent trajectory factors are estimated. In this way the mean structure of the data is determined entirely by the means of the latent trajectory factors.

Even if we focus on continuous observations, the model can be easily extended to deal with noncontinuous responses. In the literature several approaches have been developed for conducting latent variable analysis with discrete data. The most popular one used in SEM is the Underlying Variable Approach (UVA), which assumes that the manifest variables are indirect observations of multivariate normal variables. As shown in Chap. 2, another main approach is to set the model within the

Generalized Linear Latent Variable Model (GLLVM) (Moustaki and Knott, 2000). It allows to link the latent variables to observed ones of different types. Although the two approaches seem different because of different assumptions, Bartholomew and Knott (1999) have noticed an equivalence between them. Hence, the choice of one method over the other for practical applications depends both on the properties of the estimates and on the efficiency of the estimation procedures.

4.2.1 Estimation

As in the classical SEM approach, model estimation is obtained by minimizing a fitting function depending on the discrepancy between the theoretical covariance matrix of the observed variables, Σ, and the correspondent sample covariance matrix, S. Hence, the information coming from the data is considered to be sufficient to get a unique estimation value of the parameters, that is the model is identifiable (Bollen, 1989). In this case, the inclusion of the information coming from the mean structure μ is also required.

Specifically, from Eqs. (4.4) and (4.5) we define μ and Σ as

$$\mu = \Lambda(\tau + \Gamma \bar{w}) \tag{4.6}$$

$$\Sigma = \begin{bmatrix} \Lambda(\Gamma S_{ww}\Gamma' + \Psi)\Lambda' + \Theta_\varepsilon & \Lambda\Gamma S_{ww} \\ S_{ww}\Gamma'\Lambda' & S_{ww} \end{bmatrix} \tag{4.7}$$

where $\bar{w}$ is the sample mean vector of the covariates and S_{ww} the corresponding covariance matrix. Different fitting functions can be chosen according to the nature of the vector y. The following Maximum Likelihood fitting function can be used (Muthén and Khoo, 1998) if it is either a multivariate normally distributed or its components do not present excessive kurtosis,

$$F_{ML} = \ln|\Sigma| - \ln|S| + tr(\Sigma^{-1}S) - p + (\bar{y} - \mu)'\Sigma^{-1}(\bar{y} - \mu). \tag{4.8}$$

As well known, under regular conditions, F_{ML} has desirable asymptotic properties as it gives asymptotically efficient estimators of the parameters and associated correct test statistics. For evaluating the goodness of fit of the SEM models the most used statistic is defined as $(n-1)F_{ML}$. It is asymptotically distributed as a chi-square with degrees of freedom equal to the number of the variances and covariances in Σ minus the number of the estimated parameters.

4.3 Modelling nonlinearity in latent curve models

Regardless of the specific form, there are many situations in which a linear model of change does not correspond to the theoretical model of interest. Trends over time often follow a nonlinear trajectory, and ignoring nonlinearity can lead to misspecified models and biased inferences.

Two different kinds of nonlinearity can be observed: *(i)* nonlinearity of *form*, which refers to the nonlinearity of the trajectory function; *(ii)* nonlinearity in the *parameters*, that is in the subject-specific coefficients. This second type of nonlinearity is problematic in latent curve models, because LCM is a confirmatory factor model in which parameters enter linearly.

Several approaches have been developed in order to overcome these limitations. Some define an a priori formal function, and evaluate its goodness of fit to the observed data. Polynomial trajectories are commonly applied, but also bounded curves can be taken into account. On the other hand, other methods do not make assumptions on the global form of the trajectory function. They fit curves that are more reflective of the characteristic of the given data set. On this regard we shall discuss completely latent and linear piecewise trajectory models.

4.3.1 Polynomial trajectories

A straightforward formulation of a nonlinear model which is linear in the parameters is based on polynomial functions, such as quadratic or cubic growth curves. Nonlinearity is often modeled as follows

$$y_{it} = \sum_{p=0}^{P} \beta_{ip}(t-1)^p + \varepsilon_{it} \quad i=1,...,n \quad t=1,...,T_i. \tag{4.9}$$

The linear model corresponds to a first degree ($p=1$) polynomial trajectory. Given an adequate number of repeated measures to allow for proper model identification, higher-order polynomials can be specified as regression coefficients in the Λ matrix. For example, for an unconditional quadratic model, given $\eta = (\beta_{i0}, \beta_{i1}, \beta_{i2})^T$, the factor loading matrix results

$$\Lambda = \begin{pmatrix} 1 & 0 & 0 \\ 1 & 1 & 1 \\ 1 & 2 & 4 \\ \vdots & \vdots & \vdots \\ 1 & T-1 & (T-1)^2 \end{pmatrix} \tag{4.10}$$

One necessary condition for model identification is that we must have at least as many known to be identified parameters as we have unknown parameters. In general, a polynomial function of degree p requires a minimum of $T = p+2$ repeated observations (Bollen and Curran, 2006).

The number of latent variables included in η is dependent on the trend that is modeled. While the intercept is typically interpreted as initial status, the interpretation of the slope depends on the number of parameter in η. For example, consider the quadratic LCM. The intercept continues to reflect the model-implied value at the initial assessment, whereas the linear component of the quadratic model describes the instantaneous rate of change at the initial assessment. However, as this number increases, the interpretation becomes more difficult.

4.3.2 Exponential trajectories

Polynomial functions are not always well suited to capture all forms of nonlinear change. Social and psychological processes often change at a nonlinear rate with a tendency for performance to stabilize after a period of time. An application of a quadratic function to such data might do well in capturing the increase in ability at the beginning, but would then imply that the ability decreases thereafter. Although many interesting nonlinear functions are available (e.g. exponential, logistic, Gompertz), not all are easy to incorporate directly into LCM since the hypothesized model has to be linear in the parameters.

Du Toit and Cudeck (2001) describe methods for the estimation of several types of nonlinear functions within the SEM framework particularly focusing on the exponential model of change. Exponential functions are common to model growth or decay that tends toward an asymptote. The exponential trajectory is defined as

$$y_{it} = \beta_{i0} + \beta_{i1}[1 - \exp(-\gamma(t-1))] + \varepsilon_{it}, \quad t = 1,...,T \quad i = 1,..,n, \quad (4.11)$$

where β_{i0} represents the intercept of the trajectory at the initial time period, since for $t = 1$ $\exp(-\gamma 0) = 1$, β_{i1} represents the model-implied expected total change in y as time goes to infinity. Finally, γ affects the exponential rate of change in y over time.

Du Toit and Cudeck (2001) demonstrate that within the SEM framework we can estimate random components for the intercept and slope, but we need to treat γ as fixed. This implies that individual variability is allowed in the initial starting point and in the total amount of change over time, but the rate of change is fixed for all cases in the sample.

The factor loadings for this exponential curve have a more complex form than polynomial models, that is

$$\Lambda = \begin{pmatrix} 1 & 0 \\ 1 & 1 - \exp(-\gamma(1)) \\ \vdots & \vdots \\ 1 & 1 - \exp(-\gamma(T-1)) \end{pmatrix} \quad (4.12)$$

The constraints on the factor loadings require a nonlinear function that depends on γ. These nonlinear constraints can either be estimated directly, as in Eq. (4.12), or using a higher-order polynomial (see Du Toit and Cudeck, 2001). One advantage of this exponential trajectory is that unlike the polynomial family of functions, the exponential one is strictly monotonic. That is, the exponential trajectory approaches an asymptote without changing directions. In contrast, for example, the quadratic polynomial trajectory will change directions at an inflection point and continue toward plus or minus infinity. Since most social and psychological processes are bounded, there is an advantage to use a function that does not increase or decrease without limits.

Other methods have been developed for estimating exponential functions within the SEM framework (see e.g. Browne and Du Toit (1991) and Browne (1993)), but we refer to the Du Toit and Cudeck (2001) since it is the most commonly applied.

4.3.3 Complex nonlinear curves

There are empirical situations where complex nonlinear trajectories of change can make difficult a priori specification of the elements in η. Instead of fitting predefined trajectories, there are alternative methods in which we can fit functions that are more reflective of the characteristic of the given empirical data set. In this section, we illustrate complete latent models and piecewise linear trajectories.

4.3.3.1 Complete latent models

The *completely latent* model introduced by Meredith and Tisak (1990) is one of the most often applied method. Given a two-factor model, it consists in modelling curvilinear trajectories by freeing one or more of the loadings in the latent curve model. As suggested by Aber and McArdle (1991), the approach "stretches" the unit of time. It allows the value of time to be estimated in order to linearize the relation between time and y_{it} by transforming the metric of time. Meredith and Tisak (1990) proposed to set the first loading to zero, the second to 1 to define the metric of the latent factor, and to freely estimate λ_t for $t = 3,...,T$.
Hence, Eq. (4.4) results

$$\begin{pmatrix} y_{i1} \\ y_{i2} \\ y_{i3} \\ \vdots \\ y_{iT} \end{pmatrix} = \begin{pmatrix} 1 & 0 \\ 1 & 1 \\ 1 & \lambda_3 \\ \vdots & \vdots \\ 1 & \lambda_T \end{pmatrix} \begin{pmatrix} \beta_{i0} \\ \beta_{i1} \end{pmatrix} + \begin{pmatrix} \varepsilon_{i1} \\ \varepsilon_{i2} \\ \varepsilon_{i3} \\ \vdots \\ \varepsilon_{iT} \end{pmatrix} \qquad i = 1,...,n. \qquad (4.13)$$

where y_{it} is the value of the response variable y for the individual i at time point t, $i = 1,...,n$ and $t = 1,...,T$, whereas $\varepsilon_{it}, i = 1,...,n, \quad t = 1,...,T$ are the disturbances assumed to be normally distributed with zero means and heteroscedastic variances. The estimated loadings $\lambda_t, t = 2,...,T$, will reflect the observed change between times 1 and t. On the other hand, McArdle (1988) suggested to fix $\lambda_1 = 0$ and $\lambda_T = 1$, and to freely estimate all of the loadings between the first and last time point.

Although the overall model fit statistics will be identical in the two formulations, the estimated loadings have a different interpretation. Specifically, in the latter, the free loadings will reflect the proportion of change between two time points relative to the total change occurring from the first to the last time point.

The free loadings give flexibility in fitting nonlinear forms, and the model identification is less demanding than in the polynomial approach (Bollen and Curran, 2006).

4.3.3.2 Piecewise linear trajectories

Another method for modelling nonlinear relations over time is to approximate the nonlinear function through the use of two or more piecewise linear polynomial or splines. This model comes from the mixed modelling framework (e.g. Snijders and Bosker, 1999), but has been less widely utilized in the SEM approach.

The general procedure is to identify a fixed transition point during the time period under study, and to fit a linear trajectory up to that point and a linear trajectory after it. The transition point might be explicit and determined theoretically, or it may be more data driven.

Although recent works have explored the ability to model transition points that vary randomly over individuals, these methods are currently not well developed. We thus focus on fitting a piecewise linear latent curve model to repeated measures in which all individuals share the same transition point. The subject specific trajectory is thus

$$y_{it} = \beta_{i0} + \beta_{i1}\lambda_{1t} + \beta_{i2}\lambda_{2t} + \varepsilon_{it}, \qquad (4.14)$$

where λ_{1t} represents one value of time at assessment t, and λ_{2t} represents a second value of time at assessment t. It is with these two time values that we will be able to combine the two linear trajectories. Conceptually, the first piece will bring the trajectory up to the transition point t^* and then allow the second piece to continue after the transition. Thus, the factor loading matrix will be

$$\Lambda = \begin{pmatrix} 1 & 0 & 0 \\ \vdots & \vdots & \vdots \\ 1 & \lambda_{1t^*} & 0 \\ 1 & \lambda_{1t^*} & 1 \\ \vdots & \vdots & \vdots \\ 1 & \lambda_{1t^*} & \lambda_{2T} \end{pmatrix}$$

where the first column of Λ represents the intercept factor, the second column represents the first linear piece, and the third column represents the second linear piece. The coding of time nicely highlights that the first piece is defining the trajectory up to the transition point, after which the first piece turns over to the trajectory to the second piece. Similarly, the second piece makes no contribution to the trajectory prior to the transition, but picks up the trajectory after that point.

4 Nonlinearity in the analysis of longitudinal data

As before, because the intercept and two linear pieces are treated as random variables, these can be expressed as

$$\beta_{i0} = \mu_{\beta_0} + \zeta_{\beta_{i0}}$$
$$\beta_{i1} = \mu_{\beta_1} + \zeta_{\beta_{i1}}$$
$$\beta_{i2} = \mu_{\beta_2} + \zeta_{\beta_{i2}}.$$

The interpretation of the model parameters is straightforward. The means of the three latent factors represent the fixed effects for the corresponding trajectory components. On the other hand, the variance estimates of each factor represent the degree of individual variability around each of the fixed effects. The covariances among the latent factors represent linear relations among the intercept and two linear pieces. An interesting parameter in this model is the covariance between the first and second linear pieces. This covariance represents the association between individual differences in rates of change prior to and following the transition point. In terms of identification, Bollen and Curran (2006) showed that a piecewise linear LCM is identified with five or more waves of data.

4.4 Case study: analysis of university student achievements

This section examines how several methods used to model nonlinearity in longitudinal observations perform on real data sets. Longitudinal analyses of university student data allow to fulfil the new requirements of the university system. With the beginning of the Bologna process, a series of reforms are initiated with the aim to create an European Higher Education area, in which students could choose from a wide and transparent range of high quality courses and benefit from smooth recognition procedures. Hence, the evaluation of formative processes has received a growing attention by policy makers and public agents in order to identify critical factors for achievement that can improve curricula, instructional strategies, and conditions for learning.

An important emerging problem is the comparison between students' performances: *(i)* when different supporting and tutoring actions are adopted during the course of studies, and *(ii)* in presence of different personal situations.

4.4.1 The data

The data set analysed was extracted from the Data WareHouse (DWH) of the University of Bologna. This latter is a system that collects and constantly updates information by integrating data coming from sources of different nature. The project started in 2002 in order to support planning, control and decision processes.

Table 4.1 Descriptive statistics for GRAD1, GRAD2, and NOGRAD

	GRAD1		GRAD2		NOGRAD	
	Mean	StDev	Mean	StDev	Mean	StDev
y_1	7.62	1.49	5.77	1.86	3.87	1.70
y_2	8.51	1.67	6.35	1.76	4.53	1.85
y_3	9.81	1.63	7.61	2.00	4.31	1.98
y_4	–	–	4.51	2.47	4.14	2.16
y_5	–	–	–	–	3.05	1.83

Table 4.2 Correlation matrices for GRAD1, GRAD2, and NOGRAD

	GRAD1			GRAD2				NOGRAD				
y_1	1.00			1.00				1.00				
y_2	−0.07	1.00		−0.11	1.00			0.25	1.00			
y_3	−0.31	−0.48	1.00	0.02	−0.14	1.00		0.15	0.23	1.00		
y_4	–	–	–	−0.44	−0.33	−0.54	1.00	−0.02	0.21	0.10	1.00	
y_5	–	–	–	–	–	–	–	0.01	−0.14	−0.02	−0.05	1.00

The DWH contains a great amount of information per each student and allows to build the overall university student career. It is also possible to find socio-demographic information (gender, country/region of origin, etc.) and the mark obtained in the final exam of the High School.

In this chapter, we analyze a cohort of $n = 714$ students enrolled in 2001 at the Faculty of Economics of the University of Bologna. Five different time points (academic years) are observed: $t_1 = 2001/2002, t_2 = 2002/2003, t_3 = 2003/2004, t_4 = 2004/2005, t_5 = 2005/2006$. Bianconcini et al. (2007) show that within the cohort it is possible to distinguish three different patterns:

1. Students ($n_1 = 195$) who graduated in t_3 (GRAD1).
2. Students ($n_2 = 268$) who graduated in t_4 (GRAD2).
3. Students ($n_3 = 251$) who have not graduated yet (NOGRAD).

Only the first group graduates on time. To build an indicator of the student performance we decided to involve the two most relevant variables present in the DWH, that is, the mark (ranging from 18 to 30 cum laude), and the number of credits associated to each exam (ranging from 2 to 15). In more detail, the response variable y_{it} is computed as the weighted average mark obtained by each student i ($i = 1, 2, ..., 714$) over time t_l ($l = 1–5$) divided by the total number of credits required to get the degree, equal to 160. The weights are given by the credits corresponding to each exam. Thus the variable obtained is continuous and it can range from 0, if the student does not take any exam, to a maximum that depends on both the number of credits expected in each academic year, and on the average of the marks.

Tables 4.1 and 4.2 show for each pattern, the means, the standard deviations, and the correlation matrices of the response variable across time.

We can observe that GRAD1 is the group of students that presents the best average performance, increasing almost linearly during the 3 years. The performance of the students belonging to GRAD2 is quite good in the first three years, but decreases suddenly in the last year. It may be due to the fact that students prefer to achieve their degree of study in the year t_4 despite the marks obtained or, more commonly, they almost conclude their studies in the first three years, and spend the last one in preparing few exams and the final thesis. The group of NOGRAD shows a low average performance in all the time points observed. As for the correlation values, they are quite low for all the groups indicating that in general there are no strong associations between lagged performance indicators. Only for GRAD2, y_4 presents negative correlations with y_1, y_2 and y_3. It confirms the different behaviour of this group of students in t_4.

4.4.2 Results

Before fitting latent curve models to the data it is necessary to evaluate if GRAD1, GRAD2 and NOGRAD can be considered as three samples of the same population. Indeed, since the three groups have three time points in common we can test if any difference in these observed points can be ignored. On this regard, the data not observed for GRAD2 and NOGRAD are considered missing by design. Hence, a missing data three group analysis can be conducted by assuming that the three groups have been drawn from a single population (Muthén and Khoo, 1998). The tested model is the one where equality constraints are imposed for mean vector and covariance matrix elements that the three cohorts have in common. Bianconcini et al. (2007) showed that this hypothesis is rejected for these data, and, hence, the three patterns cannot be considered random samples from the same population. This requires a different latent growth specification for each of the three cohorts. Latent curve analyses are implemented by using LISREL 8.8.

Concerning the students who got their degree at t_1, only three time points are available. Hence, only a linear growth model with uncorrelated residuals among the achievement scores is identified. Since the main aim of the analysis is to show how to deal with nonlinearity in longitudinal data, we refer the reader to Bianconcini et al. (2007) for a complete study of the temporal pattern of GRAD1 students.

4.4.2.1 The GRAD2 students

The linear growth model for students who got their degree at t_4 does not fit well. The source of this misfit should not be only sought in the covariance structure, but also in the mean structure. This suggests that a linear growth assumption is not realistic, so we explored nonlinear trajectories. As shown in the previous section, several approaches can be followed in order to estimate nonlinear LCM. We started by consider the simplest extension of a linear LCM, and fitted a polynomial trajectory model. Since we have four waves of observations, a quadratic polynomial is correctly identified. The results are reported in Table 4.3. The model fits the data poorly, as indicated by the chi-square value with one degree of freedom. Hence, we need to estimate a different nonlinear model for these data. Since it is difficult to

Table 4.3 Parameter estimation for the unconditional nonlinear models for GRAD2 students

	Quadratic		Completely latent	
	Estimate	SE	Estimate	SE
λ_1	–	–	–	–
λ_2	–	–	−0.499	0.189
λ_3	–	–	−1.423	0.304
λ_4	–	–	–	–
λ_5	–	–	–	–
Mean				
β_{i0}	5.41	0.11	5.77	0.11
β_{i1}	2.21	0.19	−1.28	0.22
β_{i2}	−0.77	0.07	–	–
Variance				
β_{i0}	4.92	1.24	1.04	0.16
β_{i1}	8.93	2.30	0.29	0.79
β_{i2}	0.84	0.23	–	–
Covariances				
β_{i0},β_{i1}	6.15	1.51	−0.87	0.22
β_{i0},β_{i2}	−1.75	0.39	–	–
β_{i1},β_{i2}	2.59	0.69	–	–
χ^2	55.45		2.19	
df	1		1	
p-value	0.000		0.14	

define a priori a functional form for the GRAD2 data, we apply the complete latent approach by following the time parametrization suggested by McArdle (1988). We only fix $\lambda_1 = 0$ and $\lambda_4 = 1$, and all others are freely estimated. By correlating the residuals between t_2 and t_3, and t_3 and t_4, the model fit results are excellent (Table 4.3), according to the chi-square statistic [chi-squared: 2.19, df: 1, p-value: 0.14]. The values of the freely estimated loadings are $\hat{\lambda}_2 = -0.50$, and $\hat{\lambda}_3 = -1.42$, revealing the nonlinear pattern observed in the means. In particular, -0.50 reflects the decreasing change between t_1 and t_2 relative to the total change occurring from the first to the last time point. On the other hand, -1.42 is the decreasing change between t_2 and t_3 relative to the total change occurred in the entire period of time of 4 years.

The means corresponding to the intercept ($\hat{\mu}_{\beta_0} = 5.767$) and slope ($\hat{\mu}_{\beta_1} = -1.277$) factors are both significant, as well as the variance of the intercept ($\hat{\psi}_{\beta_0} = 1.036$), but not of the slope ($\hat{\psi}_{\beta_1} = 0.293$). These variance components reflect that there are individual differences in the starting point, but not in the nonlinear rate of change over time. On the other hand, there is significant negative covariance between the random intercept and slope, equal to -0.866. Although a linear interpretation cannot be given to these results, the value of the intercept indicates that the initial level of the analyzed cohort is quite good. On the other hand, the negative slope estimate as well as the negative loading estimates imply an increasing growth in its performance over time and a steeply decrease at the end of the studies.

4.4.2.2 The NOGRAD students

For the students who did not graduated yet, we have observed the score for five time points. Also in this case, a linear growth model fits the data poorly, whereas a quadratic trend with correlated residuals is adequate (Bianconcini et al., 2007). This is confirmed by the chi-square value (Table 4.4) corresponding to a quadratic trend with two degrees of freedom equal to 3.82 ($p = 0.15$). All the mean estimates are

Table 4.4 Parameter estimation for the unconditional nonlinear models for NOGRAD students

	Quadratic		Completely latent	
	Estimate	SE	Estimate	SE
λ_1	–	–	–	–
λ_2	–	–	−0.77	0.22
λ_3	–	–	−0.39	0.18
λ_4	–	–	−0.24	0.17
λ_5	–	–	–	–
Mean				
β_{i0}	3.90	0.11	3.90	0.10
β_{i1}	0.77	0.13	−0.86	0.14
β_{i2}	−0.24	0.03	–	–
Variance				
β_{i0}	1.79	0.58	0.31	0.11
β_{i1}	2.90	0.85	0.87	0.37
β_{i2}	0.18	0.05	–	–
Covariances				
β_{i0},β_{i1}	−1.19	0.64	−0.32	0.15
β_{i0},β_{i2}	0.19	0.13	–	–
β_{i1},β_{i2}	−0.70	0.18	–	–
χ^2	3.82		8.94	
df	2		7	
p-value	0.15		0.26	

significant and equal to 3.90, 0.77, and −0.24 for β_{i0}, β_{i1}, and β_{i2}, respectively. The variance estimates are all significant ($\hat{\psi}_{\beta_0} = 1.79, \hat{\psi}_{\beta_1} = 2.90, \hat{\psi}_{\beta_2} = 0.18$) and also the covariance between β_{i1} and β_{i2}, equal to −0.70.

Thus the students belonging to this cohort present a low initial level and, although the value β_{i1} indicates on average a positive linear growth, the negative value of β_{i2} highlights that its increment decreases over time. Taken jointly, these results reflect that the performance of this group of students is in general lower than GRAD1 and GRAD2.

We can also estimate a complete latent model for this group of data, without making any assumption on the functional form of the latent curve. As in the previous case, we fix $\lambda_1 = 0$ and $\lambda_4 = 1$, and all others are freely estimated. The results are reported in Table 4.4. We found that this model fits well the data, with a chi-square statistic with seven degrees of freedom equal to 8.94 (*p*-value 0.26). The values of the freely estimated loadings are $\hat{\lambda}_2 = -0.77, \hat{\lambda}_3 = -0.39, \hat{\lambda}_4 = -0.24$, revealing the nonlinear pattern observed in the means. In particular, the relative change with respect to the total change occurring from the first to the last time points is greater between t_1 and t_2 and decreases constantly over time. The loading $\hat{\lambda}_4$ is not significant, indicating that the worst performance is in the first two academic years.
The means corresponding to the intercept ($\hat{\mu}_{\beta_0} = 3.90$) and slope ($\hat{\mu}_{\beta_1} = -0.86$) factors are both significant, as well as the variance for the intercept ($\hat{\psi}_{\beta_0} = 2.87$), but not for the slope ($\hat{\psi}_{\beta_1} = 2.36$). These variance components reflect there are individual differences in the starting point, and also in the nonlinear rate of change over time. On the other hand, there is significant negative covariance between the random intercept and slope, equal to -2.04. Hence, there is a similar variability in both the initial status and rate of growth for this group of students, and the relationship among the two growth components is negative. Negative are also the loading estimates which imply a slightly increasing growth in its performance over time.

4.5 Concluding remarks

This chapter discussed the analysis of longitudinal data focusing on several methods developed for modelling nonlinear latent curves in the SEM framework. Based on the idea that individuals differ in their growth over time, and they are likely to have different temporal behaviours as a function of differences in particular characteristics, several methods have been developed with the main goal of treating nonlinear dynamics via models which are linear in the parameters. An application study on a cohort of students enrolled in 2001 at the University of Bologna showed how different approaches provide good estimates of the nonlinear growth. In particular, models based on complete latent trajectories always seem to fit well the nonlinear data, but when the functional form of the trajectory is known a priori, as in the case of the NOGRAD student pattern, this approach is to be preferred relative to the fully latent model.

Chapter 5
Multilevel models for the evaluation of educational institutions: a review

Leonardo Grilli and Carla Rampichini

5.1 The evaluation of educational institutions

The methodology for the evaluation of educational systems is being developed in different fields, such as educational statistics, psychometrics, sociology and econometrics. Each discipline has developed approaches suitable for the analysis of particular aspects of the evaluation process. For example, educational statistics focuses on learning curves using standardized scores, while econometrics mainly deals with private returns (e.g. in terms of wages) or social returns (e.g. in terms of productivity). Anyway, there is a considerable overlap among the fields, for example peer effects are studied both in educational statistics, as a major topic, and econometrics, as a minor topic.

In this review we focus on the methods for comparing educational institutions. Most of the literature concerns primary and secondary schools rather than universities. This preponderance is due to several factors: (*i*) primary and secondary education is compulsory and has an enormous social and economic impact; (*ii*) the majority of schools are under the responsibility of a single subject, namely the State; (*iii*) the schools share a core curriculum in mathematics and reading that allow to build standardized tests. Indeed, the potentialities of standardized tests attracted much methodological work. Anyway, most of the topics we consider in this review apply to both schools and universities and we often make reference to the evaluation of universities, which is our own research area.

The interest in the evaluation of the educational system is proved by some recent special issues edited by top-level scientific journals: Journal of Econometrics (The

---

Leonardo Grilli
Department of Statistics "G. Parenti", University of Florence, Viale Morgagni 59, 50134, Florence, Italy, e-mail: grilli@ds.unifi.it

Carla Rampichini
Department of Statistics "G. Parenti", University of Florence, Viale Morgagni 59, 50134, Florence, Italy, e-mail: rampichini@ds.unifi.it

econometrics of higher education, Lawrence and Marsh (2004)), Journal of Educational and Behavioral Statistics (Value-Added Assessment, Wainer (2004), Journal of the Royal Statistical Society series A (Performance monitoring in the public services, Bird (2004)).

The research activity in the evaluation of educational institutions influences, and is influenced by, the real applications in school districts or states. For example, in UK the publication of the ranking of schools based on raw measures of achievement (the so called *league tables*) started a debate and a methodological work that induced the government to add adjusted measures (Goldstein and Spiegelhalter, 1996; Leckie and Goldstein, 2009).

As for Italy, in the last years there are have been many research projects on the evaluation of universities (Chiandotto et al., 2005; Boero and Staffolani, 2006; Fabbris, 2007; Capursi and Ghellini, 2008), in addition to the intense activity of the National Committee for the Evaluation of the University System (www.cnvsu.it). On the other hand, the evaluation of primary and secondary schools played a minor role, even if there have been several projects financed by the government (www.invalsi.it) and some regions (e.g., Lombardia: www.irrelombardia.it). In Italy the research is more on universities than on schools mainly because a law imposed the evaluation of the universities since 1993. Indeed, in Italy the standardized tests on student achievement are few and occasional and the Government is still working to build the evaluation system. Anyway, the availability of the standardized tests of the OECD-PISA surveys is generating some valuable research activity (Martini and Ricci, 2007).

This review is written from a statistician's point of view, so the focus is on the methodological challenges connected with statistical modelling and data analysis. The second Section is devoted to the definition of effectiveness in education, while the third Section deals with multilevel models and their role in assessing effectiveness. The fourth Section gathers several statistical issues arising in effectiveness evaluation, while the fifth Section discusses the use of model results. The sixth Section concludes with some remarks.

5.2 Effectiveness

The effectiveness of an organization is the degree of achievement of its institutional targets. In the case of education (schools, universities) some targets are *internal*, such as the attainment of an adequate level of knowledge, while other targets are *external*, such as a high proportion of employed graduates or a good consistency between job and curriculum.

The degree of achievement of the targets can be measured in absolute terms (*absolute effectiveness* or *impact analysis*) or in relative terms (*relative* or *comparative effectiveness*). Absolute effectiveness is appropriate for the evaluation of interventions, e.g. a specific vocational training course, while relative effectiveness is suited for situations with many institutions offering the same service and thus interest focuses on comparing institutions. In a comparative setting, the effectiveness is usually operationalized as a measure of performance *adjusted* for the factors out of the control of the institution. In other words, the effectiveness is seen as an extra-performance entirely due to the behaviour of the institution itself.

In terms of economic theory, the issue of comparative effectiveness can be viewed through the Principal-Agent-User model (Fabbri et al., 1996). In the context of education, the Principal is the Ministry of Education, the Agents are the educational institutions (schools or universities) and the Users are the students. The subjects are in a situation of asymmetric information and need some kind of assessment of the service offered by the Agents: in fact, each User has to choose one Agent (the best for her), while the Principal wishes to rank the Agents in terms of effectiveness in order to understand the good practices and to take actions to improve effectiveness (e.g. assigning incentives).

The key point is that the quality of the output of the educational process cannot be defined in absolute terms, but only with respect to the effects on the students. However, the effects on the students are affected by the features of the students themselves, so if two institutions of similar quality have students with markedly different degrees of motivation and ability, the outcome of the two institutions is likely to be quite different. Therefore, a fair comparison of educational institutions requires to control for the characteristics of the students, in other words education is a field where the evaluation of the Agents must be adjusted for the features of the Users. In economic terms, the customers (students) are also inputs of the production function of the educational institution.

The educational process leads to multiple outcomes, so many measures of effectiveness are conceivable. As for the university, relevant internal measures are the drop-out rate, the duration of studies (time to the degree), the number of credits after a given period and the satisfaction of the students expressed through questionnaires; relevant external measures are the occupational status at a certain date after degree (employed or not), the duration of unemployment (time to first job), the wage, the job satisfaction and the consistency between job and curriculum. The definition of the outcome to be studied depends on the purpose of the evaluation and ultimately on the policy objectives. Since the stake-holders (government, management, students) give different weights to the outcomes according to their preferences, the evaluation system should avoid summarizing the various kinds of effectiveness into a single overall indicator.

In general the effectiveness is a feature that is outcome and time specific. Therefore judgements about schools need to address at least five key questions: (*i*) Effective in promoting which outcomes? (*ii*) Effective over what period of time? (*iii*) Effective for whom? (*iv*) Effective for which curriculum stage? (*v*) Effective in what educational policy or regional context?

A framework for the assessment of school effectiveness is outlined by Hanushek (1986). A broad review of the methodological and statistical issues connected with performance indicators is given by Bird et al. (2005).

5.2.1 The value-added approach

The analysis of the educational process is difficult, so the quality of educational institutions is usually measured via a *value-added* approach, where the process is a black-box and the output, called *outcome*, is evaluated in the light of the input. In this perspective, the effectiveness is just the value added by the school:

[value added] = [actual outcome] − [expected outcome given the input]

Empirical studies have found that the differences in student outcomes across schools are due mainly to differences in student prior achievement and socio-economic background and for a minor part to differences in school factors such as teachers ability, organization and so on. Thus comparing the unadjusted outcomes is markedly unfair and a value-added approach is needed.

There is an extensive literature on *value-added* student achievement, where the main methodological point is how to properly adjust the final raw achievement for the initial conditions (initial level of knowledge, motivation, socio-economic status, etc.). As explained by Tekwe et al. (2004), "Value-Added is a term used to label methods of assessment of school/teacher performance that measure the knowledge gained by individual students from 1 year to the next and then use that measure as the basis for a performance assessment system. It can be used more generally to refer to any method of assessment that adjusts for a valid measure of incoming knowledge or ability".

The issue of adjustment is crucial also in external effectiveness evaluations (employment chances, consistency between job and curriculum), but in such cases the adjustment for a *ceteris paribus* comparison is even more difficult: in fact, there is no initial measure of the outcome under study and the external nature of the result requires adjusting also for external conditions (e.g. the unemployment rate). In essence, to achieve a fair evaluation the main difficulty is to make a proper adjustment.

5.2.2 Type A and type B effectiveness

The kind of adjustment required for assessing effectiveness is not the same for the various subjects interested in the results. In this regard, it is useful to distinguish between two types of effectiveness. In fact, a potential student (User) and the Ministry of Education (Principal) are interested in different types of effectiveness of the educational institutions (Agents):

- Type A – Potential student: interested in comparing the results she can obtain by enrolling in different institutions, irrespective of the way such results are yielded;
- Type B – Ministry of Education: interested in assessing the "production process" in order to evaluate the ability of the institutions to exploit the available resources.

5 Multilevel models for the evaluation of educational institutions: a review   65

The two types of effectiveness are called A and B after Raudenbush and Willms (1995), who focused on internal measures, but the concept naturally extends to external measures. In a comparative setting, the effectiveness is usually assessed through a measure of performance *adjusted* for the factors out of the control of the institution, so the difference between Type A and Type B effectiveness simply lies in the kind of adjustment:

- Type A effectiveness: performance of the Agent adjusted for the features of its Users;
- Type B effectiveness: performance of the Agent adjusted for the features of its Users, the features of the Agent itself (out of its control) and the context in which it operates.

In the evaluation of schools or universities the features of the students to adjust for are the initial knowledge, ability, motivation etc., or proxies easier to measure, such as the socio-economic status. Examples of features of the institutions to adjust for are the public or private status, the student/teacher ratio and the amount of funding. The features of the context requiring adjustment depend on the kind of evaluation, for example to assess the effectiveness in terms of chances of employment an adjustment should be made for the conditions of the local labour market.

As pointed out by Raudenbush and Willms (1995), in practice the adjustment required for the assessment of Type B effectiveness is particularly difficult, as it involves many variables whose measurement is problematic.

5.3 Multilevel models as a tool for measuring effectiveness

The statistical models for assessing the relative effectiveness of educational institutions must face two main problems:

- adjustment: a fair comparison requires to adjust the raw outcome for several factors, depending on the type of effectiveness;
- quantification of uncertainty (accidental variability): this is necessary in order to formulate judgements supported by empirical evidence, accounting for sampling variability and other sources of error, such as fluctuations in the unobserved features of the institutions and measurement error.

The raw rankings, sometimes called *league tables*, ignore both issues (Goldstein and Spiegelhalter, 1996; Goldstein and Leckie, 2008; Leckie and Goldstein, 2009).

The main statistical tool for making a proper adjustment, while quantifying uncertainty, is regression. However, in a comparative evaluation of educational institutions, standard regression models (such as the Generalized Linear Models) are not adequate as they do not take into account a crucial feature of the data, namely the hierarchical structure. In fact, the students are nested into the institutions and the aim is to measure the effectiveness of the institutions using outcomes defined at the student level. From a statistical viewpoint, standard regression models make unsuitable assumptions on the variance-covariance structure since they assume independence of the observations, while the results of the students of the same institution are positively correlated as they share several unobserved factors at the institution level. The consequence is a poor quantification of uncertainty (and in nonlinear models also a systematic attenuation of the regression coefficients). In addition, standard models are unable to represent some key features, e.g. varying slopes.

A class of models well suited for assessing the relative effectiveness of institutions is that of *multilevel* models, also known as *mixed* models or *hierarchical* models. The reason is that multilevel models allow to:

- specify distinct sub-models for the behaviour of the institutions and the behaviour of their users;
- represent adequately the variance-covariance structure, achieving a good quantification of the uncertainty;
- represent explicitly the concept of effectiveness by means of a random effect added to the linear predictor.

There are plenty of textbooks on multilevel modelling. Snijders and Bosker (1999) is an excellent introduction. Hox (2002) has fewer details, but it covers a wider range of topics. Raudenbush and Bryk (2002) present the models in a careful way along with thoroughly discussed applications. Goldstein (2003) is a classical, though not easy, reference with wide coverage and many educational applications. A useful handbook is De Leeuw and Meijer (2008).

The web is rich of resources on multilevel modelling, for example the Centre for Multilevel Modelling at www.cmm.bristol.ac.uk. There is also a very active email discussion group for exchanging information and suggestions about multilevel modelling (see www.jiscmail.ac.uk/lists/multilevel.html).

Multilevel models can be fitted with Maximum Likelihood or Bayesian methods (Raudenbush and Bryk, 2002; Goldstein, 2003), using specialized software (e.g. MLwiN, HLM) or procedures of statistical packages such as SAS, Stata, R, Mplus.

5.3.1 *The random intercept model*

The basic multilevel model is the *linear random intercept model*:

$$y_{ij} = \alpha + \beta \mathbf{x}_{ij} + \gamma \mathbf{w}_j + u_j + e_{ij} \tag{5.1}$$

where j indexes the level 2 units (clusters) and i indexes the level 1 units (subjects). In terms of the Principal-Agent-User model outlined in the previous Section, the clusters are the Agents and the subjects are the Users. Specifically, in the evaluation of schools the clusters are the schools and the subjects are the students.

The variables in the model are: y_{ij}, the outcome of student i of school j (a raw measure of performance); $\mathbf{x}_{ij}$, a vector with the features of student i of school j; $\mathbf{w}_j$, a vector with the features of school j and the context in which it operates.

5 Multilevel models for the evaluation of educational institutions: a review

Moreover, u_j is the random effect of school j, i.e. an unobservable quantity characterizing such a school and shared by all its students. The term u_j is an adjusted measure of performance: in fact, it is a residual component that captures all the relevant factors at the school level not accounted for by the covariates and thus its meaning depends on which covariates enter the model. The effect u_j is called "random" because it is a random variable, assuming independence among the schools. For consistency of the estimates, the crucial assumption on u_j is that its expectation conditionally on the covariates is null (*exogeneity*). Less crucial, but standard assumptions are the *homoscedasticity*, i.e. the u_j have constant variance σ_u^2, and the *normality* of the distribution.

Finally, the level 1 errors e_{ij} are residual components taking into account all the unobserved factors at the student level making the outcome y_{ij} different from what predicted by the covariates and the random effect. The e_{ij} are assumed independent among students and independent of u_j. The other standard assumptions are similar to those on u_j, i.e. exogeneity, homoscedasticity (with variance denoted as σ_e^2) and normality.

The model is named random intercept since each school has its own intercept $\alpha + \gamma \mathbf{w}_j + u_j$ that has both fixed and random components. However, the slopes are assumed to be constant across schools, so the regression lines are parallel (see the left panel of Fig. 5.1).

To make clear the value-added interpretation, model (5.1) can be written as follows:

$$y_{ij} - \left(\alpha + \beta \mathbf{x}_{ij} + \gamma \mathbf{w}_j\right) = u_j + e_{ij} \qquad (5.2)$$

actual outcome − expected outcome = value added + residual

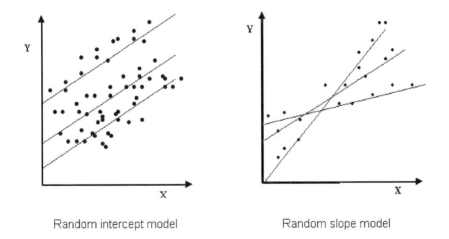

Random intercept model   Random slope model

Fig. 5.1 Regression lines in a random intercept model and in a random slope model.

The expected outcome is the outcome predicted by the model on the basis of the available school-level and student-level covariates. For student i of school j the difference between actual and expected outcome has a school-level component u_j (the value added) and a residual student-level component e_{ij}. The value added u_j is thus a school-level unexplained deviation of the actual outcome from the expected outcome. Since what is expected depends on the covariates, the meaning of the value-added term depends on how the model adjusts for the covariates, namely: (*i*) which covariates are included in the model; (*ii*) how the covariates enter the model (nonlinearities, interactions).

To illustrate the random intercept model, let us consider a simple example where y_{ij} is a measure of final achievement and the only available covariate is the corresponding measure of prior achievement x_{ij}. While x_{ij} is a student-level covariate, its school mean $\bar{x}_j$ is a school-level covariate measuring the quality of the context. In educational research the slope of $\bar{x}_j$, called *contextual coefficient*, is often found to be significant, meaning that the context has an effect on the individual outcomes. For example, assume the contextual effect is positive and consider two students i and k with the same prior achievement: if the school attended by student k has a higher mean prior achievement then the school attended by student i, then the model predicts an higher final achievement for student k. The reason is that the school attended by student k operates in a more favourable context that substantially improves the learning process.

In order to allow for contextual effects, the random intercept model for a single covariate should be specified as:

$$\begin{aligned} y_{ij} &= \alpha + \beta x_{ij} + \gamma \bar{x}_j + u_j + e_{ij} \\ &= (\alpha + \gamma \bar{x}_j + u_j) + \beta x_{ij} + e_{ij} \end{aligned} \tag{5.3}$$

where the covariate (prior achievement) has a *within slope* β and a *contextual slope* γ. Indeed, when the contextual coefficient is not null the covariate has a within slope different from the between slope and also from the total slope, see Snijders and Bosker (1999) and Raudenbush and Bryk (2002).

In model (5.3) all the school factors beyond the cluster mean of the covariate (school mean of prior achievement) are included in the random effect u_j which is broadly interpreted as the effect of school practice or value added. Denoting with A_{ij} and B_{ij} the Type A and B effects for student i of school j outlined in the previous Section, model (5.3) implies

$$A_{ij} = \gamma \bar{x}_j + u_j \tag{5.4}$$

$$B_{ij} = u_j \tag{5.5}$$

Thus the random intercept model implies *uniform* Type A and B effects, i.e. attending a given school has the same effect for all the students, regardless of their features. In statistical terms, there is no interaction between school practice and student features. The uniformity of the effects leads to straightforward rankings of the schools: once the model is fitted, the schools can be ranked on the basis of the estimated Type

A or Type B effects, yielding two rankings that may differ in a substantial way if contextual effects are relevant.

5.3.2 The random slope model

Unfortunately, uniform effects are often a restrictive assumption since typically a given school practice has more or less impact on student learning depending on the kind of student under consideration. Some schools are egalitarian, trying to reduce the gap in the prior achievement, while other schools are competitive, tending to boost the initial differences: in statistical terms, competitive schools have an higher slope on prior achievement. A multilevel model accounting for varying slopes is the *linear random slope model*:

$$y_{ij} = \alpha + \beta \mathbf{x}_{ij} + \gamma \mathbf{w}_j + u_{0j} + \mathbf{u}_{1j} \mathbf{z}_{ij} + e_{ij} \quad (5.6)$$

where $\mathbf{z}_{ij}$ is the subset of student-level covariates $\mathbf{x}_{ij}$ having a random slope.

The random slope extension of model (5.3) is

$$\begin{aligned} y_{ij} &= \alpha + \beta x_{ij} + \gamma \bar{x}_j + u_{0j} + u_{1j} x_{ij} + e_{ij} \\ &= (\alpha + \gamma \bar{x}_j + u_{0j}) + (\beta + u_{1j}) x_{ij} + e_{ij} \end{aligned} \quad (5.7)$$

so each school has its own regression line as depicted in the right panel of Fig. 5.1. Snijders and Bosker (1999, Sect. 5.3.1) discuss some specification issues.

Model (5.7) implies the following Type A and B effects:

$$A_{ij} = \gamma \bar{x}_j + u_{0j} + u_{1j} x_{ij} \quad (5.8)$$

$$B_{ij} = u_{0j} + u_{1j} x_{ij} \quad (5.9)$$

Thus the random slope model implies *non-uniform* school effects, i.e. attending a given school does not have the same effect for all the students, since the effect depends on the features of the student under consideration. Non-uniform effects make difficult to rank the schools, since the ranking changes whenever two regression lines cross, and any couple of schools has a different crossing point. A practical solution is to use the covariates to define a few relevant *profiles* (e.g. low-achievement student, medium-achievement student etc.) and to produce one ranking for each profile.

5.3.3 Cross-level interactions

As in standard regression, multilevel models can be extended to allow for interactions, namely the effect of a covariate depends on the level of another covariate. However, in multilevel analysis there is a special kind of interaction that is important for a fine modelling of the relationships between hierarchical levels: it is the *cross-level interaction*, i.e. the interaction between an individual-level covariate and a cluster-level covariate. For example, Ladd and Walsh (2002) used non parametric regression to show that the effect of the prior school mean score on the final individual score depends on the prior individual score: most students benefit from being in a school with high-scoring schoolmates, but for students with a very low prior score the effect is reversed. This situation can be modelled through cross-level interactions, for example model (5.3) may be expanded as

$$y_{ij} = \alpha + \beta x_{ij} + \gamma \bar{x}_j + \delta x_{ij} \bar{x}_j + u_j + e_{ij} \qquad (5.10)$$

so the effect of the prior school mean score $\bar{x}_j$ is $\gamma + \delta x_{ij}$, which depends linearly on the prior individual score x_{ij}.

Model (5.10) implies the following Type A and B effects:

$$A_{ij} = (\gamma + \delta x_{ij}) \bar{x}_j + u_j \qquad (5.11)$$

$$B_{ij} = u_j \qquad (5.12)$$

Thus, the random intercept model with cross-level interactions implies uniform Type B school effects, but *non-uniform* Type A school effects. Thus, there is a unique ranking of schools based on Type B effects, while the ranking based on Type A effects depends on student features.

5.3.4 Fixed versus random effects

As implied by the name, the random intercept and random slope models treat the cluster residual effects as random variables. Alternatively, such effects could be treated as unknown fixed quantities, i.e. fixed effects.

In principle, the choice between fixed and random effects is straightforward: use fixed effects whenever you wish to make inference on the clusters in the data; use random effects whenever you wish to make inference on a population of clusters, assuming that the clusters in the data are a random sample from such a population (Snijders and Berkhof, 2008). However, the choice is complicated by other, more practical, considerations. In fact, fixed effects models have the advantage of requiring fewer assumptions: there is no need to specify the distribution of the random effects, nor to assume that they are uncorrelated with the covariates (exogeneity). Moreover, Draper and Gittoes (2004) demonstrate the large sample functional equivalence between a method based on indirect standardization and a fixed effects model. Also note that when the cluster sizes become large, the fixed parameter estimates yielded by a random effects model tend towards the fixed parameter estimates yielded by a fixed effects model. See, for example, Wooldridge (2002).

Unfortunately, fixed effects models are unable to include cluster-level covariates: the technical reason is that cluster-level covariates would be perfectly collinear with the cluster indicators, while an intuitive explanation is that the fixed effects fully explain the cluster-level variability, so there is no scope for cluster-level explanatory variables. In a value-added analysis, the impossibility to include the covariates of the school/context is a serious limitation. For example, Type B effects cannot be estimated with a fixed effects model.

Another drawback of the fixed effects approach is the incidental parameter problem arising in non-linear models, yielding inconsistent estimators of all the parameters. Wooldridge (2002) gives some details.

The random effects approach is generally to be preferred, even if it entails a risk of misspecification of the conditional distribution of the random effects given the covariates, yielding biased inferences. It is therefore crucial to check the assumptions on the random effects and possibly adopt alternative specifications: for example, the non-normality of the random effects can be addressed by using a discrete distribution with estimable support points and masses (yielding non-parametric maximum likelihood estimates), while the correlation of the random effects with the covariates (endogeneity) can be solved by extending the model with the cluster means (Snijders and Bosker, 1999).

A further advantage of a random effects model is the availability of empirical Bayes (shrunken) residuals to predict the cluster effects, as discussed in Sect. 5.5.1.

5.3.5 Non-linear and multivariate multilevel models

The nature of the outcome determines the kind of multilevel (mixed) model to be used:

- continuous (test score, wage ...): linear mixed model;
- count (number of enrolled students ...): e.g. Poisson mixed model;
- time (time to degree, time to get first job ...): duration mixed model;
- binary (dropout, employment status ...): e.g. logistic mixed model;
- ordinal (satisfaction, grade ...): e.g proportional odds mixed model;
- nominal (type of job, course subject ...): e.g multinomial logit mixed model.

All the previous models belong to the class of Generalized Linear Mixed Models (GLMM). A wider class including also Rasch, IRT, factor and structural equation models is the Generalized Linear Latent And Mixed Models (GLLAMM: Skrondal and Rabe-Hesketh (2004)). The GLLAMM framework is not only a relevant theoretical advance, but it also gives the researchers an easy way to extend and integrate their models: for example, in the GLLAMM framework it is quite straightforward to specify and fit multilevel IRT models.

To give an idea of the applications of non-linear and/or multivariate multilevel models, we mention some of our works on graduates' placement using data from the Italian system: multilevel discrete-time survival models for the time to obtain the first job (Biggeri et al., 2001; Grilli, 2005); multilevel chain graph models to study the probability of employment after one year (Gottard et al., 2007); multilevel factor models for ordinal indicators to study the satisfaction on several aspects of the current job (Grilli and Rampichini, 2007a); multilevel multinomial logit models for studying where the skills needed for the current job have been acquired (Grilli and Rampichini, 2007b).

5.3.6 *Multilevel models for non-hierarchical structures*

The multilevel models considered so far are appropriate only for hierarchical (also called nested) structures. Two extensions are worth to mention: models for cross-classified structures and models for multiple membership.

A structure is called cross-classified when the individuals are classified along two or more dimensions. For example, students may be classified by school and neighborhood, so the model has both school random effects and neighborhood random effects. See Raudenbush (1993) and Rasbash and Goldstein (1994).

A multiple membership multilevel model takes into account that some individuals may change their cluster. For example, during a school cycle of 5 years a student may spend 4 years in school A and then move to school B, where she takes a final examination aimed at assessing the learning during the whole cycle. It is clearly unfair to ascribe the gain of such a student only to school B, as in a standard multilevel model. A more reasonable assumption is that the gain of such a student is due to school A for 4/5 and to school B for 1/5, as in a multiple membership model.

An example of multiple membership model is the Layered Mixed Effects Model (Sanders and Horn, 1994), which has the limitation of not allowing covariates. However, Browne et al. (2001) show that the multiple membership feature can be added to any multilevel model, encompassing models with covariates and crossed random effects. Goldstein et al. (2007) and Leckie (2009) apply multiple membership and cross-classified models to the analysis of pupil achievement.

5.4 Issues in model specification

The implementation of statistical models for value-added analysis raises several questions, where statistical issues and policy considerations are often inextricably mixed. The following considerations hold in general for value-added models, regardless of their multilevel nature.

5.4.1 *Simple versus complex models*

The value-added approach recognizes that the learning process is too complex to be fully modelled, so the pragmatic aim of accountability is pursued. Therefore, it is recommended to keep the model as simple as possible. As noted by Tekwe et al. (2004), "there is a natural desire on the part of the public and the educational establishment that implementation of school accountability systems involve simple methods understood by many, not just those with extensive methodological training". The authors state that simple models are to be preferred if they are "just as good as" complex models, so there is a "burden of proof" on value-added measures developed from complex models.

Tekwe et al. (2004) made an empirical comparison of several value-added models using data from a medium sized Florida district of 22 elementary schools. They found that the rankings originated from different models were highly correlated, with the notable exception of a mixed model with socio-economic covariates at the student level. Models without covariates produced similar rankings, e.g. a simple fixed effects model on the change score yielded essentially the same results as a complex Layered Mixed Effects Model. So the main question is if and how a value-added model should adjust for student-level and school-level covariates.

5.4.2 To adjust or not to adjust?

The notion of value-added implies to adjust the final achievement at least for the prior achievement. It is widely accepted that "the minimal requirement for valid institutional comparison is an analysis based on individual level data which adjusts for intake differences" (Aitkin and Longford, 1986).

Unfortunately, there is no general agreement on which other factors should be controlled for. The value-added measures should be purged from the factors out of the control of the school, but in practice the separation between factors under control and factors out of control is not so clean. Tekwe et al. (2004) state that "if schools are partly but not wholly responsible for the effects of covariates, then bias results from either including or excluding them". Usually the schools are not responsible for the socio-economic status (SES) of their students, which is mainly determined by the features of the district where the school is located, so adjusting for SES is appropriate. However, if the admission to the schools is selective, it may be that the worst schools have few students with high SES just because they are known to be bad. In that case the adjustment is unduly beneficial for the bad schools.

The decision to adjust for socio-economic factors also depends on the purpose of the evaluation process. Tekwe et al. (2004) stated that a model that adjusts only for the prior achievement "... might be preferred in a low-stakes accountability system that provides incentives and resources for 'less effective' schools to improve and that does not base salary raises on the value-added measures. In a high stakes system, however, where teachers' salaries and school budgets depend on 'high performance', not adjusting for significant socio-demographic factors could encourage the flight of good teachers and administrators from schools with high percentages of poor or minority students. On the other hand, adjusting for these factors could institutionalize low expectations for poor or minority students and thereby limit their opportunity to achieve their full potential."

Ladd and Walsh (2002, pp. 3–4) discuss the issue of adjusting for race, as implemented in the value-added system of Dallas: "The educational logic for including race is not transparent. At best it serves as a proxy for income and family characteristics, such as low income and single parent families, for which other data were not available or were incomplete. In contrast, the political logic for Dallas to control for the student's race in the equation was very clear. Dallas officials wanted to make sure that schools serving minority students had the same probability of being judged an effective school as any other school. The problem is that by applying this criterion of perceived fair treatment, Dallas officials could well have been concealing some true differences in the relative effectiveness of schools serving minorities. Policy makers in other states have specifically chosen not to control for the race of the student based on political considerations of a different sort. If they were to include race as a control variable, they faced the possibility that they might be misinterpreted as sending a signal that the academic expectations for minority children are lower than those for white children. Such a message would be inconsistent with the rhetoric that underlies much of the outcomes oriented reform efforts, namely that all children can learn to high levels. While this concern about a specific demographic variable applies most pointedly to a student's race, it applies as well to other background characteristics of students, such as family income."

As statisticians, we believe that a model for value-added analysis should control for all the relevant factors, paying attention to issues such as endogeneity and measurement error. Once a good model has been fitted, how to use the results is a policy matter: for example, one can decide to publish only Type A effects.

5.4.3 Endogeneity

The difficulty of adjusting for student covariates can be seen as stemming from the correlation between such covariates and the school effects, i.e. endogeneity (Braun and Wainer, 2007). Note, however, that adding the cluster mean makes a student-level covariate uncorrelated with the school effects, so valid estimates of the Type A effects can be obtained. In a sense, the bias induced by the correlation is shifted to Type B effects. In general, the estimation of Type B effects is biased by the endogeneity of school-level covariates (Raudenbush and Willms, 1995). For example, if the less effective schools receive more resources (e.g. measured by expenditure per pupil and pupil-teacher ratio) then the estimated resource effects are attenuated due to endogeneity. Steele et al. (2007) try to solve the endogeneity problem by using a multilevel model with two simultaneous equations, one for pupil achievement and the other one for school resources.

The most common source of endogeneity is the omission of relevant covariates (Kim and Frees, 2007). Other sources are sample selection bias (Grilli and Rampichini, 2007c) and measurement error (see Sect. 5.4.5).

5.4.4 Modelling the achievement progress

The value-added accountability systems typically produce databases reporting, for any student and any subject, the measures of achievement (scores) at several grades. The measurements of achievement at different grades should be on the same scale (*vertical scaling*), which is problematic since the content of a subject varies across grades. Braun and Wainer (2007) discuss the case of mathematics, where the role of geometry in the curriculum substantially increases in later grades, so in a sense "math is not math".

When the prior achievement $y_{t-1,ij}$ is measured on the same scale as the final achievement $y_{t,ij}$, the response variable of a value-added model can be either the final achievement $y_{t,ij}$ or the difference between final and prior achievement $y_{t,ij} - y_{t-1,ij}$ (*progress*). Consider the following random intercept model on the final achievement:

$$y_{t,ij} = \alpha + \beta y_{t-1,ij} + \gamma x_{ij} + u_j + e_{ij} \qquad (5.13)$$

where x_{ij} is a student-level covariate. In this model, if $\beta = 1$ the progress, i.e. the final achievement minus the prior achievement, does not depend on the prior achievement, but only on the covariates; if $\beta < 1$ the progress is higher for students with lower prior achievement; on the contrary if $\beta > 1$ the progress is higher for students with higher prior achievement. Usually $\beta < 1$ in public schools, where one of the main goals is to reduce differences among students' abilities. Usually the behaviour of the schools in this respect is not the same, i.e. β can vary between schools, calling for a random slope model such as (5.7).

The interpretation of the parameters of model (5.13) is clearer if we subtract the prior achievement $y_{t-1,ij}$ from both sides, obtaining the corresponding random intercept model for the *progress*:

$$y_{t,ij} - y_{t-1,ij} = \alpha + (\beta - 1)y_{t-1,ij} + \gamma x_{ij} + u_j + e_{ij} \qquad (5.14)$$

The only difference in the slopes of models (5.13) and (5.14) concerns the slope of prior achievement. The slopes of the other covariates are unchanged, making clear that in model (5.13) the slopes of the covariates have to be interpreted as effects on the progress and not on the level: indeed, even if the response is the final achievement, the prior achievement is controlled for. Note that a model for the progress without adjusting for the prior achievement amounts to assume $\beta = 1$, which is in general not plausible.

In principle, the effect of the covariate x_{ij} on the final achievement $y_{t,ij}$ can be decomposed in the sum of two components: the effect on the prior achievement $y_{t-1,ij}$ and the effect on the progress $y_{t,ij} - y_{t-1,ij}$. Denoting with φ the slope of the regression of $y_{t-1,ij}$ on x_{ij}, the total effect of x_{ij} on $y_{t,ij}$ is $\varphi\beta + \gamma$. If the covariate has a cumulative effect on the achievement, then $\varphi\beta$ is likely to be greater than γ, and such a gap tends to increase with the educational grade. In applied work many value-added models omit the covariates: the rationale is that the covariates affect the achievement level but not the progress. However, such an assumption should be tested whenever possible by adding the covariates to the model.

5.4.5 Measurement error

Value-added models are based on measures of pupil achievement, usually obtained through standardized tests. The score of a test is a fallible measure of the true achievement, with a measurement error that depends on the reliability of the test. When the score is the response variable of the model, its measurement error is captured by the model error and there are no consequences on the estimates. However, the prior score is often used as a covariate in value-added models, as in (5.13) and (5.14), causing measurement error bias. The problem disappears in the change score model (5.14) if the prior score is omitted, but such an omission as to be tested.

For illustration, let us consider model (5.13), where the final achievement is the response and the prior achievement enters as a covariate:

$$\begin{cases} y_{t,ij} = \alpha + \beta y_{t-1,ij} + \gamma x_{ij} + u_j + e_{ij} \\ s_{t-1,ij} = y_{t-1,ij} + m_{t-1,ij} \\ s_{t,ij} = y_{t,ij} + m_{t,ij} \end{cases} \quad (5.15)$$

where the m's are measurement errors with zero mean and variance $\sigma^2_{m_{t-1}}$ for prior scores and $\sigma^2_{m_t}$ for final scores. The measurement errors are assumed to be independent of the model variables and independent across students and across occasions. Replacing the final achievement with the final score, the model becomes

$$\begin{cases} s_{t,ij} = \alpha + \beta y_{t-1,ij} + \gamma x_{ij} + u_j + e^*_{ij} \\ s_{t-1,ij} = y_{t-1,ij} + m_{t-1,ij} \end{cases} \quad (5.16)$$

where $e^*_{ij} = e_{ij} + m_{t,ij}$. Under the standard assumptions, e^*_{ij} is independent of the covariates and thus β and γ can be estimated without bias. However, replacing the prior achievement with the prior score yields a student-level error correlated with the prior score and thus the estimators are biased. In model (5.16), if $\gamma = 0$ and u_j is dropped, the probability limit of the least squares estimator is $\beta \lambda_{t-1}$, where $\lambda_{t-1} = \sigma^2_{y_{t-1}} / \left(\sigma^2_{y_{t-1}} + \sigma^2_{m_{t-1}} \right)$ is the reliability of the prior score. Since the reliability is less than one, the slope of the prior achievement is biased toward zero (attenuated). Therefore, the effect of the prior achievement in not fully controlled for, so the analysis penalizes the schools with disadvantaged students. For example, in the application of Ladd and Walsh (2002) two-fifths of the differentially favourable outcome for schools serving advantaged students result from measurement error bias, so after correcting for measurement error the relative rankings of the schools change substantially.

If the model has a covariate x_{ij}, the slope of the prior achievement $y_{t-1,ij}$ is still attenuated, while the slope of x_{ij} is inflated or attenuated depending on the correlation structure. Indeed, the attenuation of the slope of $y_{t-1,ij}$ implies that the effect of $y_{t-1,ij}$ is partially controlled for, so part of its effect is absorbed by the slope of x_{ij}. For example, if x_{ij} is the socio-economic status, which typically has a positive effect on achievement, the measurement error on the prior achievement causes an upward bias in the estimation of the slope of x_{ij}.

The basic difficulty with a model whose covariates are affected by measurement error is that the model parameters are not identified, so the correction usually relies on unverifiable assumptions or on external data (Fuller, 1987). The issue of measurement error in multilevel models is discussed in Battauz et al. (2005) and in Ferrão and Goldstein (2009).

5.5 Use of the model results

Once a suitable model is fitted, the results can be used to:

1. study the relationship between the outcome and the explanatory variables;
2. rank the schools according to their effectiveness;
3. predict the outcome for a given student in a given school.

The first aim is common to all statistical models when they are used to understand real phenomena. In general, the findings can be legitimately interpreted in terms of associations since, apart from the rare controlled experiments, a casual interpretation requires strong untestable assumptions. On this point, Rubin et al. (2004) argue that "without 'heroic assumptions' causal inferences cannot be legitimately drawn". See also Raudenbush (2004), Braun and Wainer (2007), Hong and Raudenbush (2008) and Jin and Rubin (2009).

5.5.1 Ranking the schools

The aim of ranking the schools according to effectiveness has two main purposes: accountability and information to the potential users. The rankings are widely used for accountability, especially as a tool to identify schools with anomalous performances deserving special attention. On the other hand, the dissemination of the rankings to the potential users is less frequent and it is still the object of heated debates (Ladd and Walsh, 2002; Goldstein and Leckie, 2008; Leckie and Goldstein, 2009).

School rankings are derived from school-level residuals, which can be seen as predictions of the random effects representing the effectiveness. The residuals can be obtained in two main ways (Raudenbush and Willms, 1995): in a conventional way by subtracting the expected outcome from the observed outcome, or via empirical Bayes (EB). The conventional method gives unbiased estimates of the school effects, while the EB method produces the so called *shrunken residuals*, which are biased but efficient estimates of the school effects. The shrinkage pulls the conventional residual towards its population mean, i.e. zero, depending on the cluster size: the smaller the size the greater the shrinkage. Apart from efficiency considerations, shrunken residuals are usually preferred in school evaluation settings, since they provide protection against the fortuitous assignment of a school to the top or bottom of the ranking.

Since the residuals are affected by the sampling variability and other sources of error, the corresponding ranking has a degree of uncertainty. Such uncertainty is difficult to represent, since it involves multiple comparisons. The usual approach is to build pairwise confidence intervals (Goldstein and Healy, 1995), even if more sophisticated approaches are possible (Afshartous and Wolf, 2007). For example, Fig. 5.2 reports the EB predictions of random effects along with 95% pairwise bars for a set of schools: the effectiveness of two schools is statistically different whenever the 95% pairwise bars of the two residuals do not overlap.

Figure 5.2 is a typical picture arising in empirical analyses: only a few schools at the top and at the bottom of the ranking are statistically different, so there is little evidence for ranking the schools. In addition, Leckie and Goldstein (2009) point out that potential students are interested in future rather than past effectiveness: this implies larger error bars around the residuals, so the comparisons are even more inconclusive.

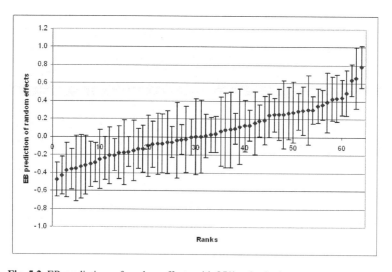

Fig. 5.2 EB predictions of random effects with 95% pairwise bars.

5.5.2 Predicting the outcome

The prediction of the outcome for a given student is relevant for guidance purposes. After estimation of the parameters and prediction of the random effects, it is possible to predict the outcome of a student with certain features in a specific school. For example, the predicted outcome from the random intercept model (5.3) is:

$$\hat{y}_{ij} = (\hat{\alpha} + \hat{\gamma}\bar{x}_j + \hat{u}_j) + \hat{\beta}x_{ij} \tag{5.17}$$

Since the random intercept model without cross-level interactions implies uniform school effects, the ranking of the schools based on the predicted outcomes (5.17) is the same as the ranking based on the predicted Type A effects (5.4).

The predicted outcome from the random slope model (5.7) is:

$$\hat{y}_{ij} = \left(\hat{\alpha} + \hat{\gamma}\bar{x}_j + \hat{u}_{0j}\right) + \left(\hat{\beta} + \hat{u}_{1j}\right) x_{ij} \tag{5.18}$$

In this model the ranking of the schools changes with the student's characteristics, so a student-specific prediction is needed for guidance purposes. The same is true for the random intercept model with cross-level interactions (5.10).

To guide the students in their choice, the government could set up a system where the student plugs in her characteristics x_{ij} and obtain the predicted outcome for every school. It is worth to note that the usefulness of the predictions depend on their precision, which is difficult to compute. Raudenbush and Willms (1995) show how to estimate the variance of Type A effects.

5.6 Concluding remarks

This paper has reviewed some methodological issues in the evaluation of school effectiveness focusing on the value-added approach and its implementation via multilevel modelling.

Even if multilevel models represent a theoretically satisfactory tool for the assessment of educational institutions, their implementation must face serious problems such as misspecification due to omitted variables, measurement error bias, and low power in ranking the institutions.

In general, the value-added approach itself suffers from some limitations: (*i*) it does not explain why a school is effective or ineffective; (*ii*) studies of school effects are quasi-experiments, so causal conclusions are questionable; (*iii*) a satisfactory adjustment for the input requires several good-quality covariates; (*iv*) measurement error in the covariates (especially prior achievement) may bias the slope estimates; (*v*) it is difficult to fully account for all the uncertainty; (*vi*) it is difficult to communicate the results to a non specialized audience.

In spite of its limitations, the value-added approach is an extremely useful tool to analyze the factors related with the student achievement and to identify outstanding students and schools, even if it needs to be used in conjunction with qualitative analysis in order to give reliable and effective indications for the accountability of schools.

Chapter 6
Multilevel mixture factor models for the evaluation of educational programs' effectiveness

Roberta Varriale and Caterina Giusti

6.1 Introduction

Factor models aim at explaining the associations among observed random variables in terms of fewer unobserved random variables, called common *factors*. When data have a hierarchical structure, multilevel mixture factor models are a powerful and flexible tool useful to correctly take into account the correlation between first-level units due to the data structure, and to evaluate the presence of latent sub-populations of units with some typical profile at different levels of the analysis.

In the Chapter, we describe the specification of a multilevel mixture factor model with continuous latent variables at the lower level of the analysis and a discrete latent variable at the higher level, focusing on some technical and applied features of the analysis. The theory will be illustrated by means of an application on the job satisfaction of the graduates of the University of Florence. The main aim of the analysis is to describe and summarize some aspects of job satisfaction measured at the individual level and, at the same time, to cluster higher level units (degree courses) in classes with some typical characteristics, in order to analyse their effectiveness.

The Chapter is organized as follows. In Sect. 6.2 we introduce the multilevel mixture factor model, and in Sect. 6.3 we collocate it in the Generalized Latent Variable framework. The details of estimation procedures and of model selection for multilevel mixture factor models are described in Sects. 6.4 and 6.5. Finally, in Sect. 6.6 we present and comment the main results of the case study on the evaluation of University effectiveness.

Roberta Varriale
Department of Methodology and Statistics, Tilburg University, Tilburg, The Netherlands;
Department of Statistics "G. Parenti", University of Florence, Florence, Italy,
e-mail: roberta.varriale@ds.unifi.it

Caterina Giusti
Department of Statistics and Mathematics Applied to Economics, University of Pisa, Via C. Ridolfi 10, 56124, Pisa, Italy, e-mail: caterina.giusti@ec.unipi.it

6.2 The multilevel mixture factor model

Factor models aim at finding a set of continuous latent variables, called *factors*, that contains the same information of a given set of observed variables (Bartholomew and Knott, 1999). One basic assumption of factor models states that the observed variables are measured on a set of independent units. This assumption is inadequate when units are nested in clusters having a hierarchical structure, sharing common environments, experiences and interactions: in these cases multilevel techniques are necessary in order to correctly take into account the correlation between first-level units due to the data structure. In this Chapter, attention is limited to datasets with two hierarchical levels, since the extension to more than two levels is conceptually straightforward.

The basic idea of a factor model adapted to deal with multilevel data is that some model parameters – indicator intercepts or thresholds and residual variances, factor loadings, factor means and variances – are allowed to differ across the *observed* groups (higher level units). These differences can be modeled including group dummies in the model, as in the multigroup (or fixed-effects) approach, or can be modeled with a multilevel factor model with continuous latent variables at all levels of the analysis by assuming that the group coefficients are random-effects coming from a particular distribution whose parameters should be estimated (Searle et al., 1992; Vermunt, 2003).

In a confirmatory perspective, the multilevel mixture factor model is a useful model to take into account the hierarchical structure of the data and to compare the observed groups of units, by evaluating the existence of unobserved subpopulations (classes) of groups with similar features with respect to the factor model parameters and overcoming the production of over-detailed information of the multigroup factor model, which estimates as many group coefficients as the groups (Vermunt, 2003).

In one-level context, the term finite mixture (McLachlan and Peel, 2000) or latent class model (Lazarsfield and Henry, 1968; Goodman, 1974) is typically used for models including only a categorical latent variable, whereas the term factor mixture model is used for models including both continuous latent variables and a categorical latent variable (Lubke and Muthén, 2005). Both models are usually applied to classify individual units into K latent classes with similar model parameters; in standard finite mixture models the clustering is based on the similarity of the observed item parameters (intercept or thresholds), in factor mixture models the clustering is based on the similarity of both the item parameters and/or the factor loadings. A discrete latent variable can also be used as a non parametric specification of a distribution of continuous latent variables (Aitkin, 1999; Vermunt and Magidson, 2005). Indeed, a finite mixture distribution results from the discretization of a continuous latent variable distribution into K probability masses π_k at mass points z_k; the nonparametric specification is so represented by a finite mixture model with the maximum number of identifiable latent classes.

Formally, a factor mixed model includes a categorical latent variable in the model with a multinomial distribution; besides the parameters of the factor model, also the parameters of the multinomial distribution have to be estimated.

6 Multilevel mixture factor models

In two-level context, finite mixture components, formally "represented" by a categorical latent variable, may be present at the lower or/and higher level. When there are mixture components at both levels of the analysis, the multilevel latent class model is obtained (Vermunt, 2003), otherwise we obtain the multilevel mixture factor model. In the Chapter, we only discuss two-level models characterized by continuous latent variables at the lower level and a categorical latent variable at the higher level. The main aims of this model are to analyse the underlying structure of the phenomenon at the lower level and, at the same time, classify higher level units in some latent classes with similar profiles.

Assume that there are J groups with a different number of individual units n_j, whose total number is equal to $N = \sum_{i=1}^{J} n_j$. For each individual, H items are observed. Conditional on the latent variables, the response model for the observed variables is a generalized linear model specified via a linear predictor, a link, and a distribution from the exponential family. Let y_{hij} denote the observed response on indicator h ($h = 1, \ldots, H$) of individual i ($i = 1, \ldots, n_j$) within group j ($j = 1, \ldots, J$) and let v_{hij} be the linear predictor of the response model. The conditional expectation of the response y_{hij} given the latent variables at different levels is "linked" to the linear predictor v_{hij} via a link function:

$$g(E(y_{hij}|\boldsymbol{\eta}_j)) = v_{hij} \tag{6.1}$$

where $\boldsymbol{\eta}_j = \left(\boldsymbol{\eta}_j^{(2)'}, \ldots, \boldsymbol{\eta}_j^{(L)'}\right)'$ represents all latent variables, $\boldsymbol{\eta}_j^{(l)} = \left(\eta_{1j}^{(l)}, \ldots, \eta_{M_l j}^{(l)}\right)'$ indicates all the latent variables varying at level l and M_l denotes the number of these latent variables. In particular, the latent variables varying at the individual and cluster level are denoted, respectively, with $\boldsymbol{\eta}_j^{(2)}$ and $\boldsymbol{\eta}_j^{(3)}$; indeed, since we are analysing models for datasets with one level of hierarchy, $l = 2, 3$. Following the conventions, these models are called two-level models: the individual units i are the level-1 units, and the group level units j are the level-2 units. If the items are treated as level-1 units, the models become three-level models with individual units at level 2 and groups at level 3.

Different distributional forms are allowed for each indicator and the choice among different link functions naturally follows from the scale types of the observed variables. In particular, while in the traditional literature different terms are used depending on the nature of both latent and observed variables (Bartholomew and Knott, 1999), in the following we will use only the general term factor models. Recent developments in computational statistics extended the use of estimation methods traditionally used for models with only continuous indicators to the analysis of models with any kind of response variables.

As an example, with continuous responses an identity link and a normal distribution are usually assumed, so (we do not use the subscript j, for simplicity):

$$y_{hi} = v_{hi} + e_{hi}$$

with $f(e) \sim N(0, \sigma^2)$; therefore, the conditional density of y_{hi} given the latent variables becomes:

$$f(y_{hi}|\boldsymbol{\eta}_j) = \sigma^{-1}\phi(v\sigma^{-1})$$

where ϕ represents the standard normal density. As another example, with ordinal responses several model specifications are possible. Let s, $s = 1,\ldots,S$ be the category of the ordinal response y_{hi}, the model for the cumulative probabilities is expressed by:

$$g[P(y_{hi} \leq s|\boldsymbol{\eta}_j)] = \alpha_s - v_{hi} \quad s = 1,\ldots,S-1 \tag{6.2}$$

where α_s with $\alpha_1 < \ldots < \alpha_{S-1}$ are the thresholds to be estimated. Typical choices of link function include the probit, logit and complementary log-log.

The two-level mixture factor model for continuous indicators and with one categorical latent variable at the highest level of analysis is:

$$y_{hij} = \mu_{hj} + \sum_{m=1}^{M_2} \lambda_{mh}^{(2)} \eta_{mij}^{(2)} + e_{hij}^{(2)} \tag{6.3}$$

$$\mu_{hj} = \sum_{k=1}^{K} \lambda_{kh}^{(3)} \eta_{kj}^{(3)} + e_{hj}^{(3)} \tag{6.4}$$

$$\eta_{mij}^{(2)} = \sum_{k=1}^{K} \beta_{km}^{(3)} \eta_{kj}^{(3)} + e_{mij}^{(2)} \tag{6.5}$$

where $\eta_{mij}^{(2)}$ denotes the mth common factor at individual level, $\lambda_{mh}^{(2)}$ represents the factor loading for factor m and item h and μ_{hj} is the item h intercept for each group j. The two terms $e_{hij}^{(2)}$ and $e_{hj}^{(3)}$ represent the item-specific errors at lower and higher level. The variable $\eta_{kj}^{(3)}$ in Eqs. (6.4) and (6.5) is an indicator variable taking value 1 if unit i belongs to latent class k of the categorical latent variable $\boldsymbol{\eta}_j^{(3)}$ and 0 otherwise, and $\lambda_{kh}^{(3)}$ and $\beta_{km}^{(3)}$ represent the coefficients for each class k. The classes are mutually exclusive and, for the identification of the model, $\sum_{k=1}^{K} \lambda_{kh}^{(3)} = 0$ or $\lambda_{1h}^{(3)} = 0$ and $\sum_{k=1}^{K} \beta_{km}^{(3)} = 0$ or $\beta_{1m}^{(3)} = 0$. The term $e_{mij}^{(2)}$ represents a residual component of the relationship between $\eta_{mij}^{(2)}$ and $\boldsymbol{\eta}_j^{(3)}$.

The variable $\boldsymbol{\eta}_j^{(3)} = \left(\eta_{1j}^{(3)},\ldots,\eta_{Kj}^{(3)}\right)$ has a multinomial distribution, with:

$$\pi_k = P\left(\eta_j^{(3)} = k\right) = P\left(\eta_{kj}^{(3)} = 1\right) = \frac{\exp(\gamma_k)}{\sum_{t=1}^{K}\exp(\gamma_t)} \tag{6.6}$$

with

$$\sum_{k=1}^{K} \pi_k = 1. \tag{6.7}$$

The term γ_k in Eq. (6.6) represents the intercept term of the linear predictor of the logit model for the expectation of the latent distribution (π_k); models with covariate effects on class membership can be defined by including covariate effects in this linear term.

6 Multilevel mixture factor models

The basic assumptions of multilevel mixture factor models are that each group belongs to no more than one latent class k, the individuals are independent inside each group conditional on the latent class k at the higher level and the H responses of individual i are independent of each other given the continuous latent variables at the individual level and the group latent class membership, which is often referred to as the local independence assumption (Bartholomew and Knott, 1999).

The $\boldsymbol{\eta}_j^{(2)}$ are usually assumed to be normally independent and identically distributed with:

$$\boldsymbol{\eta}_j^{(2)} \sim MN(\mathbf{0}, \boldsymbol{\Psi}^{(2)})$$

where MN indicates the Multivariate Normal distribution and $\boldsymbol{\Psi}^{(2)}$ is the $M_2 \times M_2$ variance and covariance matrix with elements $\psi_{mm'}^{(2)}$.

It is also assumed that the item-specific error at both levels of the analysis, $e_{hij}^{(2)}$ and $e_{hj}^{(3)}$, are mutually independent and identically normally distributed.

In the most general case of multilevel mixture factor analysis, both $\lambda_{kh}^{(3)}$ and $\beta_{km}^{(3)}$ in Eqs. (6.4) and (6.5) may differ across higher-level mixture components in order to capture the differences between individuals due to the hierarchical data structure. Two special cases of the model are obtained by constraining these terms. In the first case, $\lambda_{kh}^{(3)} = 0$, therefore the outcome variables are not directly affected by the higher level latent class and the item intercepts do not vary across group-level classes; in the second case, $\beta_{km}^{(3)} = 0$, so the individual-level latent variable does not vary across group-level classes. The first case is typically used when the researchers' interest is in classifying the higher level units and comparing the obtained groups with a confirmative approach, "pushing" up the information collected at the individual-level to the group-level through the different "steps" of the model. The second case is typically used with an exploratory approach, aiming at analysing separately the lower and higher structure of the data.

The model is represented in Fig. 6.1. Following the conventions, circles represent latent variables and rectangles represent observed variables. The latent categorical variable are indicated with a filled circle. The arrows connecting latent and/or observed variables do not necessarily represent linear relations and possible correlations among latent variables or among items are represented with dotted lines. The nested frames represent the nested levels, for example, variables located within the outer frame labeled j vary between clusters and have a j subscript (Skrondal and Rabe-Hesketh, 2004).

6.3 The Generalized Latent Variable framework

The two-level mixture factor model described so far belongs to the Generalized Latent Variable framework introduced by Muthén (2008) and Vermunt (2007). This general framework integrates specific methodologies for latent variable modelling, such as multilevel, longitudinal and structural equation models as well as item response models, factor models and so on, in a global theoretical context and allows to define models with any combination of categorical and continuous latent variables at each level of the hierarchy.

The generalized latent variable model is formally described by two elements: the response model for the observed variables conditional on the latent variables and the model for the latent variables. Using the index j to denote an independent observation corresponding to the highest level of the hierarchy, the two-level mixture factor model is expressed by:

$$g[E(\mathbf{y}_j|\boldsymbol{\eta}_j)] = \mathbf{Z}_j\boldsymbol{\beta} + \boldsymbol{\Lambda}^{(1)}\boldsymbol{\eta}_j \qquad (6.8)$$

$$h[E(\boldsymbol{\eta}_j^{(2)})] = \mathbf{X}_j\boldsymbol{\gamma} + \boldsymbol{\Lambda}^{(2)}\boldsymbol{\eta}_j^{(3)} \qquad (6.9)$$

where $\mathbf{y}_j$ denotes the response vector with elements y_{hij} representing the response to indicator h of each individual i belonging to group j.

In the two-level framework, the vector $\boldsymbol{\eta}_j = \left(\boldsymbol{\eta}_j^{(2)'}, \boldsymbol{\eta}_j^{(3)'}\right)'$ in Eq. (6.8) denotes the latent variables varying at the i-th and j-th level of the analysis affecting directly the observed responses. The vector $\boldsymbol{\eta}_j^{(3)}$ in Eq. (6.9) denotes the latent variables at the j-th level affecting the latent variables at the i-th level.

The two matrices $\mathbf{Z}_j$ and $\mathbf{X}_j$ with the corresponding coefficient vectors $\boldsymbol{\beta}$ and $\boldsymbol{\gamma}$ denote the fixed part of the model affecting, respectively, the observed items and the latent structure at level 2. Different links and distributions can be specified for different responses. The matrices $\boldsymbol{\Lambda}$, which elements do not vary depending on j, represent the factor loading matrix of the generalized latent variable model. In particular, $\boldsymbol{\Lambda}^{(1)}$ indicates the factor loading matrix relating the latent variables directly

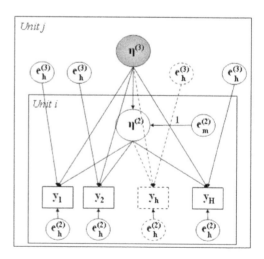

Fig. 6.1 Two-level mixture factor model.

to the outcomes and $\boldsymbol{\Lambda}^{(2)}$ indicates the factor loading matrix relating level 3 to level 2 latent variables.

Table 6.1 schematically represents different specifications of the two-level mixture factor model. In particular, a model with continuous latent variables at both levels of the analysis is called two-level factor model, while models with both continuous and categorical latent variables are called two-level mixture factor models. Which model should be selected depends on the aims of the specific research and on the substantive reason to believe in the nature, continuous or categorical, of the latent variables.

Table 6.1 Matrix of potential two-level models with underlying latent variables

	Higher level	latent variables	
Lower level latent variables	Continuous	Categorical	Combination
Continuous	A1	A2	A3
Categorical	B1	B2	B3
Combination	C1	C2	C3

Model A1, in which both the lower and higher level latent variables are continuous, is represented by the multilevel factor model, as described by Goldstein and McDonald (1988) and Longford and Muthén (1992); its extension to ordinal indicators is given by Grilli and Rampichini (2007a). Model A1 contains also three-level regression models with continuous random effects. Model B2, in which both the lower and higher level latent variables are categorical, is the multilevel latent class model. Vermunt (2003) proposes a model where lower level units are clustered based on their observed responses and higher level units are clustered based on the likelihood of their members to be in one of the unit level clusters. Vermunt (2003) also proposes a multilevel latent class model with continuous random effects at the group level (B1). Palardy and Vermunt (2009) used specification A3 to define a multilevel extension of the mixture growth model (Muthén, 2004), where two-level units are classified into homogeneous groups based on properties of their mean growth trajectories.

This brief and incomplete review of the literature shows how modelling using a combination of continuous and categorical latent variables provides an extremely general and flexible framework of analysis. Furthermore, different traditions such as growth modelling, multilevel modelling, latent class analysis are brought together using the unifying theme of latent variables.

6.4 Likelihood, estimation and posterior analysis

Recent developments in computational statistics have enhanced the feasibility of a maximum likelihood analysis in the context of multilevel mixture factor models.

In this section we briefly present the formulation of the likelihood that has to be maximized.

In two-level models, the total marginal likelihood is:

$$L(\boldsymbol{\theta}) = \prod_{j=1}^{J} L_j(\boldsymbol{\theta}) = \prod_{j=1}^{J} f^{(j)}(\mathbf{y}_{(j)}|\boldsymbol{\theta}) \qquad (6.10)$$

where L_j indicates the likelihood of group j, the groups are assumed to be independent and $\boldsymbol{\theta}$ represents the complete set of unknown parameters to be estimated. The complete likelihood can be derived recursively. In a model with $\boldsymbol{\eta}^{(2)}$ and $\boldsymbol{\eta}^{(3)}$ being, respectively, continuous latent variables at the first and second level of the analysis (not using the subscript j for the latent variables hereafter, for simplicity), the likelihood for each group j is given by:

$$L_j(\boldsymbol{\theta}) = \int_{\boldsymbol{\eta}^{(3)}} \prod_{i=1}^{n_j} L_{ij}(\boldsymbol{\theta}|\boldsymbol{\eta}^{(3)}) f(\boldsymbol{\eta}^{(3)}) d\boldsymbol{\eta}^{(3)} \qquad (6.11)$$

where the n_j level-1 units within level-2 units are assumed to be independent given the random coefficients $\boldsymbol{\eta}^{(3)}$. For each first-level unit, controlling for the effect of the latent variables at the highest level, the likelihood is expressed by:

$$L_{ij}(\boldsymbol{\theta}|\boldsymbol{\eta}^{(3)}) = \int_{\boldsymbol{\eta}^{(2)}} L_{ij}(\boldsymbol{\theta}|\boldsymbol{\eta}^{(2)},\boldsymbol{\eta}^{(3)}) f(\boldsymbol{\eta}^{(2)}|\boldsymbol{\eta}^{(3)}) d\boldsymbol{\eta}^{(2)}. \qquad (6.12)$$

Finally, considering the local independence assumption, the observed indicators are assumed to be independent given the latent variables, so:

$$L_{ij}(\boldsymbol{\theta}|\boldsymbol{\eta}^{(2)},\boldsymbol{\eta}^{(3)}) = \prod_{h=1}^{H} f(y_{hij}|\boldsymbol{\eta}^{(2)},\boldsymbol{\eta}^{(3)}) \qquad (6.13)$$

where $f(y_{hij}|\boldsymbol{\eta}^{(2)},\boldsymbol{\eta}^{(3)})$ indicates the distribution of the response variables.

When the latent variables are categorical, the multiple integrals are replaced by multiple sums. In a model with $\boldsymbol{\eta}^{(3)}$ and $\boldsymbol{\eta}^{(2)}$ being, respectively, a categorical and continuous latent variables, the likelihood is expressed by:

$$L_j(\boldsymbol{\theta}) = \sum_{k=1}^{K} P(\boldsymbol{\eta}^{(3)} = k) \prod_{i=1}^{n_j} L_{ij}(\boldsymbol{\theta}|\boldsymbol{\eta}^{(3)} = k)$$

$$L_{ij}(\boldsymbol{\theta}|\boldsymbol{\eta}^{(3)} = k) = \int_{\boldsymbol{\eta}^{(2)}} L_{ij}(\boldsymbol{\theta}|\boldsymbol{\eta}^{(2)},\boldsymbol{\eta}^{(3)} = k) f(\boldsymbol{\eta}^{(2)}|\boldsymbol{\eta}^{(3)} = k) d\boldsymbol{\eta}^{(2)}$$

$$L_{ij}(\boldsymbol{\theta}|\boldsymbol{\eta}^{(2)},\boldsymbol{\eta}^{(3)} = k) = \prod_{h=1}^{H} f(y_{hij}|\boldsymbol{\eta}^{(2)},\boldsymbol{\eta}^{(3)} = k).$$

Maximum Likelihood estimation involves finding the estimates for $\boldsymbol{\theta}$ that maximize the marginal likelihood function (or the log-likelihood function).

In maximizing the likelihood, two separated problems must be considered: solving the integrals involved in the likelihood and maximizing the likelihood function. With respect to the first aspect, while a closed form expression for these integrals is available when all responses and latent variables are continuous and normally distributed, in the other cases there are several approaches to approximating the integrals, as Laplace approximation, numerical integration using quadrature or adaptive quadrature, Monte Carlo integration (Skrondal and Rabe-Hesketh, 2004). With respect to the second aspect, several methods were proposed for maximizing the likelihood, the most common being the Expectation-Maximization (EM) algorithm and Newton-Raphson or Fisher scoring algorithms. Of course, each integration method may be combined with some maximization methods.

The main aim of a researcher using factor models is in what can be known about the latent variables after the indicators have been observed (Bartholomew and Knott, 1999). At each level of the analysis, this information is represented by the conditional density:

$$f(\boldsymbol{\eta}|\mathbf{y}) = f(\boldsymbol{\eta})f(\mathbf{y}|\boldsymbol{\eta})/f(\mathbf{y}). \tag{6.14}$$

From the point of view of social behavioral scientists, this means locating units on the dimensions of the latent space (*factor scores*), or classifying units in different *classes* representing some typical profile. Obviously, units with the same response pattern will be assigned the same factor score or class.

Some scoring methods are the ones based on the empirical Bayesian posterior distribution and the maximum likelihood method (Skrondal and Rabe-Hesketh, 2004). Usually, the firsts are the most used; indeed, while the maximum likelihood approach produces scores that are conditionally unbiased, it is not consistent with the modelling assumptions since it requires that the latent variables are considered fixed parameters and does not yield predictions for clusters with insufficient information. For this reason, we only present the two Bayesian posterior distribution methods.

With the empirical Bayesian approach, according to Bayes' theorem, the conditional posterior distribution of the latent variables given the observed variables is expressed by:

$$f(\boldsymbol{\eta}|\mathbf{y},\hat{\boldsymbol{\theta}}) = \frac{f(\mathbf{y},\boldsymbol{\eta}|\hat{\boldsymbol{\theta}})}{f(\mathbf{y}|\hat{\boldsymbol{\theta}})} = \frac{f(\mathbf{y}|\boldsymbol{\eta},\hat{\boldsymbol{\theta}})f(\boldsymbol{\eta}|\hat{\boldsymbol{\theta}})}{\int_{\boldsymbol{\eta}} f(\mathbf{y}|\boldsymbol{\eta},\hat{\boldsymbol{\theta}})f(\boldsymbol{\eta}|\hat{\boldsymbol{\theta}})} \tag{6.15}$$

where $\hat{\boldsymbol{\theta}}$ represent the estimated parameters, $f(\mathbf{y}|\hat{\boldsymbol{\theta}})$ is the distribution of the observed variables and $f(\mathbf{y},\boldsymbol{\eta}|\hat{\boldsymbol{\theta}})$ is the joint distribution of the observed and latent variables. This approach uses the term "Bayesian" since both the latent and observed variables are treated as random variables. Actually, the full Bayesian approach would assume a prior distribution for $\boldsymbol{\theta}$ in addition to the distribution for $\boldsymbol{\eta}$ and the $\boldsymbol{\theta}$ in Eq. (6.15) would be treated as fixed constants.

The computation of the posterior distribution is strictly related to the specification of the prior distribution of the latent variables. Usually, the posterior distribution cannot be expressed in closed form and heavy numerical integration is required. In factor models with continuous random variables, it follows from standard results on conditional multivariate normal densities that the posterior density is multivariate normal; for other response types, the posterior density tends to multinormality as the number of units in the clusters increases (Skrondal and Rabe-Hesketh, 2004).

After estimating the empirical Bayesian posterior distribution, two approaches can be used to estimate the factor scores (or latent class) associated to each unit: the prediction using empirical Bayes (also called a posteriori) and the prediction using empirical Bayes modal (also known as *modal* a posteriori).

The empirical Bayes prediction is the most widely used method for scoring. The predictors are represented by the mean of the posterior empirical Bayesian latent variables distribution in Eq. (6.15), so:

$$\boldsymbol{\eta}^{\text{EB}} = E(\boldsymbol{\eta}|\mathbf{y},\hat{\boldsymbol{\theta}}). \tag{6.16}$$

With continuous normal latent variables, the empirical Bayes predictor is the best linear unbiased predictor BLUP (Skrondal and Rabe-Hesketh, 2004).

The prediction using empirical Bayes modal uses the posterior mode instead of the posterior mean for the prediction of the factor scores:

$$\boldsymbol{\eta}^{\text{EBM}} = \max_{\boldsymbol{\eta}} \text{arg} \, (\boldsymbol{\eta}|\mathbf{y},\hat{\boldsymbol{\theta}}). \tag{6.17}$$

This method does not require numerical integration, so when the posterior density is approximately multivariate normal it is often used as an approximation of the empirical Bayes solutions. In particular, this method represents the standard classification method in latent class modelling since it minimize the expected misclassification rate (Skrondal and Rabe-Hesketh, 2004). Obviously, in standard factor models the predictors obtained with the empirical Bayes and empirical Bayes modal coincide.

6.5 Model selection

A number of overall and individual statistical measures of fit has been proposed in order to evaluate a specified model on the basis of empirical data. In the following, some tests based on the likelihood theory and some information criteria useful to choose between different multilevel and multilevel mixture factor models are briefly introduced.

One method to compare nested models is based on the likelihood ratio test (Agresti, 2002). However, standard asymptotic results for the test do not hold if the null hypothesis is on the boundary of the parameter space since regularity conditions would be violated; well-known examples are testing the null hypothesis relating to random effects (Self and Liang, 1987) and testing the hypothesis on the variability of the latent factors. In these cases, a rule of thumb is to divide by two the asymptotic *p*-value of the Chi-squared likelihood ratio test statistic distribution (Skrondal and

Rabe-Hesketh, 2004). Also in the mixture models framework the likelihood ratio statistic cannot be used to compare two nested models, one with k_0 classes and one with k_1 classes ($k_0 < k_1$). Indeed, under the null hypothesis of k_0 groups, some of the parameters of the model with k_1 classes lie on the boundary of the parameter space so that regularity conditions for likelihood ratio statistic to be asymptotically Chi-squared are not fulfilled. In particular, the correct null distribution of the likelihood ratio statistic is unknown (Everitt, 1988) but a lot of conjectures and simulations have been published on this topic (McLachlan and Peel, 2000).

Another approach for comparing models is based on the computation of some indexes representing a penalized form of the likelihood: as the likelihood increases with the addiction of some parameters, it is penalized by the subtraction of a term related to the number of parameters. These information criteria are generally expressed in terms of:

$$-2\log L(\boldsymbol{\theta}) + C \qquad (6.18)$$

where the first term measures the lack of fit of the model and C is the penalty term that measures the complexity of the model. The intent is therefore to choose a model to minimize this criterion.

Relating to the problem of choosing between models with different number of latent classes, a variety of textbooks and articles suggest the use of the Bayesian Information Criterion (BIC) (Schwarz, 1978) as a good indicator (Nylund et al., 2007). The BIC is expressed by:

$$BIC = -2\log L + p \times \log(N) \qquad (6.19)$$

where $\log L$ is the loglikelihood value, p is the number of parameters and N is the number of observations for the fitted model. In two-level models the number of observations can refer to both within and between levels; this distinction can make a substantial difference when determining the number of classes of a multilevel mixture model. To our knowledge, while there is a wide variety of literature available on the performance of model selection statistics for determining the number of mixture components in one-level mixture models, there are no works in the two-level context, except that of Lukočiené and Vermunt (2004). In their paper, the authors show the results of a simulation study on multilevel latent class analysis with a fixed number of classes at the lower level, aiming at individuating the best index for determining the number of mixture components at the higher level.

6.6 Case study

In this section a multilevel mixture factor model is used in order to evaluate the university external effectiveness of the degree courses of the University of Florence. As suggested by Chiandotto (2004), students' perception of the quality of the services provided by an institution can be evaluated both at the time of the degree (internal effectiveness) and some date later (external effectiveness). In particular, we evaluate the University performance from the users' subjective point of view, as perceived three years after the degree.

Different proposals on the use of multilevel methodologies to analyse both the external and internal effectiveness of the university system can be found, as some examples, in Giusti and Varriale (2008); Chiandotto et al. (in press); Chiandotto and Varriale (2006); Chiandotto and Giusti (2006). In the present application the use of multilevel mixture factor models, with a combination of continuous and categorical latent variables at different levels of the analysis, allows to fulfill two objectives, corresponding to the levels of the analysis. The "first level objective" is to understand the latent constructs underlying the phenomenon of job satisfaction using the information available at the individual level, that is the satisfaction expressed by graduated students that are employed three years after the degree. At the same time this individual information can be used to fulfill a "second level objective", to classify the study programs attended by the graduates into a small number of classes representing some typical profiles, that is to identify those programs with similar characteristics with respect to job satisfaction.

The job satisfaction is a complex process naturally considered as a latent construct not directly observable but measured by some indicators. Data come from the AlmaLaurea survey "Employment opportunities, 2005" (Almalaurea, 2006) and they concern graduates of the University of Florence. Data have a hierarchical structure, with graduates nested in different degree courses; in particular, it is interesting to investigate the effect of this level of aggregation on job satisfaction.

We consider the graduates with the old Italian university system during the summer session of the solar year 2002 who are employed at the moment of the interview, 3 years after the degree. We focus on the analysis of job satisfaction three years after the degree since it is reasonable that after that time all graduates find the job they have studied for and they are usually no more involved in specialization and training courses, except for the graduates in medicine. Obviously, as a confirmation of the results obtained with the present work, it would be interesting to repeat the same analyses when data referring to the graduates' occupational status five years after their degree will be available. For reasons of representativeness, we only consider those degree courses with at least eight employed graduates. The 1,025 graduates we include in the analysis represent almost 60% of the graduates at the University of Florence in the summer session of 2002; the total and percentage numbers of graduates in each degree course are in Table 6.2.

The questionnaire used for the Almalaurea survey "Employment opportunities" is very comprehensive, since it deals with many aspects related to the current job or the search for a job. The questionnaire section on the satisfaction with the actual job consists in 14 items. Through a correlation analysis and other preliminary considerations, we selected five of these items, measuring the satisfaction with: earnings, career opportunities, coherence with the University studies, professionalism and cultural interests. All these items are expressed on an ordinal scale with 10 categories; the items are considered as continuous variables because of the number of the categories. The average evaluation for each of the 5 items is in Table 6.3.

6 Multilevel mixture factor models

Table 6.2 Number of graduates employed three years after the degree, by degree course. Students graduated (old system degree) at the University of Florence, summer session, year 2002

Degree course	Number of employed graduates	Percentage
Architecture	216	21.07
Chemistry	9	0.88
Business economics	26	2.54
Economics	67	6.54
Philosophy	16	1.56
Law	106	10.34
Civil engineering	31	3.02
Electronic engineering	29	2.83
Mechanical engineering	23	2.24
Literature	78	7.61
Foreign lang. and literature	48	4.68
Mathematics	11	1.07
Medicine	17	1.66
Psychology	51	4.98
Biology	11	1.07
Political sciences	131	12.78
History	8	0.78
Informatics engineering	10	0.98
Environmental engineering	21	2.05
Educational sciences	102	9.95
Forest and environ. sciences	14	1.37
	1,025	100

As we can see, there are some differences between the degree courses in the mean evaluations expressed by the graduates. For example, the graduates in philosophy and history express the lowest mean evaluations for the aspects coherence and cultural interests; moreover, they give low scores to the other three aspects. At the opposite, the graduates in architecture and law are the most overall satisfied. For the graduates in medicine we observe a really high evaluation for coherence, professionalism and cultural interests, as expected, but lower mean values for career and earnings, probably because these graduates are still involved in some specialization courses. There are also some differences between similar degree courses, like the ones in engineering; for example, the interviewed who graduated in electronic engineering seem to be less satisfied with their careers with respect to their colleagues. The differences in graduates' satisfaction between degree courses show an important influence of the hierarchical data structure on job satisfaction.

Due to the results of the preliminary analyses on the correlation structure between the items and to the latent (non observable) nature of the job satisfaction, we proceeded with an exploratory (EFA) and a confirmatory one-level factor analysis (CFA). As illustrated in Sect. 6.5, likelihood ratio tests have been used to compare models with different factor loadings, while BICs have been used to compare models with different number of latent factors. In particular, with EFA we compared models with 2 and 3 latent factors measured, at the same time, by all the indicators.

Table 6.3 Mean evaluations with the selected items, by degree course. Students graduated (old system degree) at the University of Florence, summer session, year 2002

Degree course	Coherence	Professionalism	Cultural interests	Earnings	Career
Architecture	7.39	7.65	7.38	6.81	7
Chemistry	6.56	7.44	6.78	6.56	6.56
Business economics	7.85	7.85	6.85	7.19	6.92
Economics	6.91	7.34	6.46	6.85	6.78
Philosophy	4.81	6.63	5.06	5.63	5.73
Law	7.21	7.73	7.45	7.03	7.24
Civil engineering	7.74	7.68	7.29	6.55	6.41
Electronic engineering	6.55	7.14	6.83	6.28	5.97
Mechanical engineering	7.35	7.61	7.65	6.96	6.65
Literature	5.73	7.33	6.67	5.68	5.51
Foreign lang. and literature	5.71	7.15	6.4	6.08	6
Mathematics	5.36	7	6.82	6.64	5.64
Medicine	9.41	8.06	8.71	6.88	6.12
Psychology	6.2	7.12	6.75	5.59	5.82
Biology	7.82	8.55	7.18	5	5.18
Political sciences	5.53	7.16	6.57	6.29	6.43
History	3.25	7.13	5.75	6	5.75
Informatics engineering	7.4	7.2	7.1	6.3	6
Environmental engineering	7.76	8	7.38	6.9	6.45
Educational sciences	7.26	7.56	7.49	5.84	5.99
Forest and environ. sciences	6.43	7.21	7.21	5.79	5.93
	6.76	7.47	7.03	6.42	6.44

Subsequently, we run a CFA following what suggested by the correlation structure of the items and constraining to zero the loadings that resulted to be close to zero with EFA. The results of these analyses suggest the presence of two factors: one factor related to the *Cultural* features of the job, measured by career, professionalism, coherence and cultural interest, and one factor related to the *Status* of the job, measured by earnings, career and professionalism.

In order to take into account the two-level data hierarchy and to classify the degree courses in some latent classes with different profiles, we applied a two-level mixture factor model. The final model is:

$$y_{hi} = \mu_h + \sum_{m=1}^{M_2} \lambda_{mh}^{(2)} \eta_{mij}^{(2)} + e_{hij}^{(2)} \tag{6.20}$$

$$\eta_{mij}^{(2)} = \sum_{k=1}^{K} \beta_{km}^{(3)} \eta_{kj}^{(3)} + e_{mij}^{(2)} \tag{6.21}$$

At the program level, $\lambda_{kh}^{(3)} = 0$, therefore it is assumed that the degree courses differ only in the mean level of latent factors at the individual level $\left(\eta_{mij}^{(2)}\right)$ and the outcome variables are not directly affected by the higher level latent variable. In other words, $\beta_{km}^{(3)}$ represents the mean of the m-th factor at individual level for the degree courses belonging to the k-th latent class.

In the model, the items coherence and earnings are the reference items (factor loading equal to 1), respectively, for the factors *Cultural* and *Status*. At the second level of the analysis, in Eq. (6.21) $\beta_{1m}^{(3)}$ are constrained to 0 for each m, $m = 1, 2$, in order to ensure the identification.

The Bayesian Information Criterion index calculated with N equal to the number of groups is used to choose between models with different number of classes at group level. Table 6.4 shows BIC values for models composed of 1–4 classes.

Table 6.4 Two-level mixture factor model: loglikelihood and fit indexes. Students graduated (old system degree) at the University of Florence, summer session, year 2002

N classes	N param.	Log-likelihood	BIC (N obs.)	BIC (N groups)
1	18	−9775.29	19675.37	19605.38
2	21	−9750.12	19645.82	19564.17
3	24	−9737.45	19641.27	19547.97
4	27	−9733.09	19653.36	19548.38

The final two-level mixture factor model is represented in Fig. 6.2.

At the individual level, the factor structure is very similar to that found with the one-level factor analysis. Again, we acknowledge the presence of two highly correlated latent factors (Table 6.6). The first factor (*Status*) is related to the satisfaction with earnings, career and professionalism; the second factor (*Cultural*) is related to career, professionalism and to the satisfaction with cultural interests and coherence of the job with the previous studies.

Factor loadings are shown in Table 6.5. All the loadings have the same sign. As is always the case, the latent dimension underlying the global satisfaction at the program level has an arbitrary scale, which means that factor scores must be interpreted relatively to each other. The most important aspects relating to factor *Status*

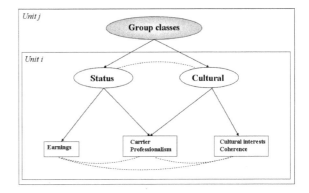

Fig. 6.2 Two-level mixture factor model. Students graduated (old system degree) at the University of Florence, summer session, year 2002.

are earnings and career, while this is the case for coherence and cultural interests with the factor *Cultural*. In other words, for each degree course, the graduates' satisfaction with the job *Status* is measured mostly by their opinion on earnings and career and the graduates' satisfaction with the job *Cultural* is measured mostly by their opinion on coherence and cultural interests. Thus, the multilevel mixture factor model gives some insides on the dimensions influencing graduates' job satisfaction at the individual level.

Table 6.5 Factor loadings. Students graduated (old system degree) at the University of Florence, summer session, year 2002

	Status	Cultural
Earnings	1	
Career	0.98	0.13
Professionalism	0.16	0.57
Coherence		1
Cultural interests		0.82

Table 6.6 Variances, covariance and *correlation* of the factors. Students graduated (old system degree) at the University of Florence, summer session, year 2002

	Status	Cultural
Status	3.23	*0.39*
Cultural	1.32	3.63

As already underlined, besides these results referring to the first level of analysis, the model expressed by (6.20) and (6.21) allows also to interpret the effect of the degree courses on graduates' job satisfaction.

At the second level of analysis, the model classifies the courses in three classes. The sizes of the three classes are different: a degree course has a probability equal to 0.45 to be in the first class, of 0.36 to be in the second one and of 0.19 in the third one (Table 6.7, last row). Due to the constraints, the class-specific effects must be interpreted in terms of deviations from the "reference class" where the effects are equal to 0; in this analysis, the reference class is the first (Fig. 6.3). The three classes differ in the mean value of the two latent factors: the second class has a higher mean level of satisfaction both for *Status* and *Cultural* and the third class has a slightly lower mean value for the factor *Status*, while the satisfaction with *Cultural* is the highest between the three.

Using the empirical Bayes modal prediction, the degree courses can be assigned to the three classes (Table 6.7, column 2), so that we can better interpret the previous results. The main part of the courses, 11 out of 21, are attributed to the reference class. For some of these courses, in particular for chemistry, informatics

6 Multilevel mixture factor models

Table 6.7 Two-level mixture factor model: study programs classification based on the empirical Bayesian posterior distribution. Students graduated (old system degree) at the University of Florence, summer session, year 2002

Degree course	Class (modal)	Prob. Class 1	Prob. Class 2	Prob. Class 3
Architecture	2	0	1	0
Chemistry	1	0.52	0.43	0.05
Business economics	2	0.01	0.99	0
Economics	2	0.14	0.86	0
Philosophy	1	1	0	0
Law	2	0	1	0
Civil engineering	2	0.01	0.68	0.3
Electronic engineering	1	0.92	0.07	0.01
Mechanical engineering	2	0.02	0.95	0.03
Literature	1	1	0	0
Foreign lang. and literature	1	1	0	0
Mathematics	1	0.89	0.1	0.01
Medicine	3	0	0.04	0.96
Psychology	1	1	0	0
Biology	3	0.01	0.01	0.98
Political sciences	1	1	0	0
History	1	0.97	0.02	0
Informatics engineering	1	0.4	0.36	0.25
Environmental engineering	2	0.01	0.87	0.11
Educational sciences	3	0	0	1
Forest and environ. sciences	1	0.61	0.15	0.24
Mean values		0.45	0.36	0.19

engineering and forest and environmental sciences, the posterior probabilities of belonging to a specific latent class at the group level are spread in the three classes (Table 6.7, columns 3 to 5). A more in-depth analysis could be useful in order to analyse the peculiarities of these courses. The degree courses belonging to the second class, the "best" for the satisfaction with both latent factors, are architecture, business economics, economics, law, civil engineering, mechanical engineering and

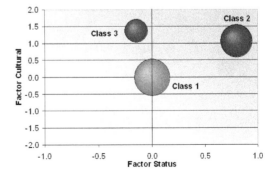

Fig. 6.3 Latent classes features (latent factors *Cultural* and *Status*. Students graduated (old system degree) at the University of Florence, summer session year 2002.

environmental engineering. Graduates in these courses developed the skills and the possibility to choose a job which guarantees a high level of satisfaction with the different aspects we considered. Graduates in the courses belonging to the third class, namely medicine, biology and educational sciences, are instead more likely to have a job with a high correspondence to their cultural interests and previous studies, while the position or status of their jobs is maybe expected to increase in the future. In particular, graduates in medicine are probably still involved in specialization and training courses, while the other graduates can also be occupied in some occasional and temporary positions because they are encountering some difficulties to find the job they studied for.

Chapter 7
A class of statistical models for evaluating services and performances

Marcella Corduas, Maria Iannario and Domenico Piccolo

7.1 Introduction

Evaluation can be described as the psychological process which a subject has to perform when a subject is requested to give a determination of merit regarding an item (the attributes of a service, a product or in general, any tangible or intangible object) using a certain ordinal scale. This process is rooted in the subject's perception of the value/quality/performance of the object under evaluation.

The mechanism governing individual choices between a set of possible alternative options has been widely studied by the latent variables theory. From a statistical point of view, however, the focus is concentrated on modelling empirical observations from sample surveys and on the investigation of the stochastic mechanism generating the ordinal data.

Sample surveys gather measures of satisfaction which are a manifest expression of respondents' constructs. For instance, measuring the satisfaction with a given service, the agreement with a specific statement, the strength of consensus on a certain rule, the perceived experience of a system's performance represent situations where a continuum latent variable (representing the profound belief of the respondent) has to be transformed by a mental process into a discrete state in order to assign an evaluation referred to the graded scale proposed by the interviewer.

Marcella Corduas
Department of Statistical Sciences, University of Naples Federico II, Via Leopoldo Rodinò, 22 I-80138, Naples, Italy, e-mail: marcella.corduas@unina.it

Maria Iannario
Department of Statistical Sciences, University of Naples Federico II, Via Leopoldo Rodinò, 22 I-80138, Naples, Italy, e-mail: maria.iannario@unina.it

Domenico Piccolo
Department of Statistical Sciences, University of Naples Federico II, Via Leopoldo Rodinò, 22 I-80138, Naples, Italy, e-mail: domenico.piccolo@unina.it

The general pattern of responses to a questionnaire aimed at evaluating a service surely presents common features originating from a few latent traits (constructs, variables, factors). This condition, of course, is not immediately recognizable from the observed ratings. Empirical evidences confirm that similarities, differences and contrasts among responses are very common. However, although a remarkable number of hypothetical patterns can be conjectured for rating distributions, only a small subset of them are observed in practice with noticeable frequency.

In the previous Chapters, various approaches widely discussed and applied in the literature have been examined. Attention has been focused on generalized linear latent models, Item Response Theory, unobserved variable approach, and several methodological developments and tools for real applications have been discussed. The main merit of such approaches relies on the possibility of dealing with manifest and latent variables starting from a unique paradigm.

In this Chapter a mixture distribution for ordinal data is introduced. This proposal, as with any innovative tool, is not aimed at replacing existing modelling which are surely based on theories widely investigated and experimented. Instead, it is intended as an additional tool which may be of help in order to better understand real data providing an alternative point of view.

The Chapter is organized as follows: firstly, a simplified description of the evaluation process is presented in order to specify the final result originated from such a process as the combined effect of two unobservable components, one related to the individual feeling for the object under evaluation and the other related to the intrinsic uncertainty which affects any human decision. Later, in Sect. 7.3, a class of models (named CUB) is logically derived from these assumptions, and properties and extensions are illustrated. In Sect. 7.4, inferential issues and numerical procedures for maximum likelihood parameter estimation and related asymptotic inference are discussed; in addition, the main steps of the EM estimation algorithm is provided for a specific CUB model. Sections 7.5 and 7.6 deal with possible applications of this class of models for ordinal data analysis. In particular, a data set concerning students' satisfaction with university "orientation" services is examined. Finally, some remarks on further generalizations and extensions conclude this contribution.

7.2 Unobserved components in the evaluation process

Perception is a cognitive process by which a subject attains awareness or understanding of sensory information and translates them into a form that is meaningful for his/her conscience. In real applications, where statistical tools are needed to analyze evaluation data a simplified archetype of such a process may be of help. In this respect, we can start by considering a simple example concerning university teaching assessment. When a student is asked to answer a specific question about the quality of teaching, he/she has to bring his/her perception of the problem into focus and then he/she has to summarize this perception into a well-defined category using a finite set of ordinal values.

7 A class of statistical models for evaluating services and performances   101

Thus, the final evaluation is the effect of complex causes. It is influenced by considerations fully related to the object of evaluation, but also by the inherent uncertainty that accompanies any human decisions. Moreover, individual behaviours may significantly differ depending on a specific subject's characteristics. Consequently, judgements can be considered as the realization of a stochastic phenomenon which needs to be modelled by taking into account the impact of individual covariates on the expression of the perception. Specifically, with respect to the assessment of university teaching and services, a sensible approach should study how expressed evaluations change with students' profiles.

For this purpose, final judgements, originating from a mental process of selecting among a discrete number of options, can be described as the compounding of two elements:

- a *primary* component, generated by the respondent's sound impression related to awareness and a full understanding of problems, personal or previous experience, group partnership, etc.;
- a *secondary* component, generated by the intrinsic uncertainty affecting the final choice. This may be due, for instance, to the amount of time spent elaborating the answer, the limited range of available information, a partial understanding of the question or to subject's laziness.

Then, the psychological mechanism, by which the choice is made, is the result of a personal *feeling* for the object under judgement and an inherent *uncertainty* associated with the selection of the ordinal value of the response.

7.2.1 Rationale for a new class of models

The interpretation of the respondents' final choice as a weighted combination of individual *feeling (agreement)* and some intrinsic *uncertainty (fuzziness)* leads to the definition of a mixture distribution that will be formally introduced in the following pages. Here, we briefly discuss the rationale behind this new probabilistic model.

Feeling is usually related to subjects' motivations, whereas *uncertainty* mostly depends on circumstances that surround the process of judging. Consequently, the first component is related to the several causes leading to a certain choice, whereas the second is simply related to the confidence/firmness/resolution of such a choice.

In order to model the first component, a *shifted Binomial* random variable is introduced. This is motivated by two arguments.

From a statistical point of view, a standard Binomial distribution is generated by adding several independent and identically distributed Bernoulli choices. Then, we may think that when a subject chooses a rating (among m possible categories) he/she excludes the others by a pairwise comparison (D'Elia, 2000, 2001). For instance, assuming a m-points graded scale (where 1 is related to the best rate), assigning the third grade to an item means that this rate is worse than the first two and better than the other $(m-3)$ ones. Generally, one chooses $(Y = y)$ when the selected choice y is not preferred to the previous $(y-1)$ but it is instead preferable to the remaining $(m-y)$ alternatives. If $(1-\xi)$ and ξ are the probability that each comparison is lost and won, respectively, a given sequence of "failure/success" has a probability of $(1-\xi)^{y-1}\xi^{m-y}$. A combinatorial argument proves immediately that the probability of a given choice is: $\binom{m-1}{y-1}(1-\xi)^{y-1}\xi^{m-y}$, for $y=1,2,\ldots,m$. Of course, this reasoning assumes that the random variables describing comparisons are both independent and identically distributed, and, as often happens for a statistical model, this provides a crude approximation of the respondents' effective behaviour.

From a heuristic point of view, the shifted Binomial distribution is able to map a continuous latent variable (characterized by a single mode distribution: Normal, Student-t, logistic, etc.) into a discrete set of values $\{1,2,\ldots,m\}$. The shape of the resulting distribution depends on the way the cut-points are originally chosen. This fact adds further flexibility in modelling the observations since it allows for very different mode location and skewness.

The second component, describing *uncertainty*, is given by a *discrete Uniform* random variable over the support $\{1,2,\ldots,m\}$. This probability distribution is intended as an extreme solution to represent the evaluation process. In this regard, we are not stating that people answer questions in a purely random manner, instead we are saying that the uncertainty affecting any choice can, at worst, be constituted by a situation where no category prevails over the others, and that is the case of a uniform distribution. In fact, the latter maximizes entropy with respect to any other distribution which shares the same finite discrete support.

The random variables related to *feeling* and *uncertainty*, are then combined in a mixture distribution with different weights (π) and $(1-\pi)$ respectively, which denote *propensities* of the subject for one or the other way of constructing his/her choice. In addition, the interpretation for the two unobserved components implies an immediate meaning for the two involved parameters: $(1-\xi)$ will be considered a measure of agreement/feeling for the item of interest whereas $(1-\pi)$ will provide a measure of fuzziness/uncertainty that accompanies the choice.

Some further remarks on the rationale behind the proposed mixture distribution may be useful at this stage. Firstly, it is important to make clear that we are not conjecturing that the population is composed of two subgroups of respondents, each behaving according to one of the two above-mentioned probability distributions.

Secondly, it is worth noticing that *uncertainty*, the component related to choosing, is completely different from *randomness*, which is instead a concept related to sampling variability of surveys.

7.3 Specification and properties of CUB models

Formally, CUB models are specified by considering the ordinal response y as a realization of a discrete random variable Y defined on the support $\{y=1,2,\ldots,m\}$. For given $m>3$, the random variable Y is a mixture of Uniform and shifted Binomial random variables and its probability mass function is given by:

7 A class of statistical models for evaluating services and performances

$$Pr(Y=y) = \pi \binom{m-1}{y-1}(1-\xi)^{y-1}\xi^{m-y} + (1-\pi)\frac{1}{m}, \quad y=1,2,\ldots,m,$$

where $\pi \in (0,1]$ and $\xi \in [0,1]$ (Piccolo, 2003; D'Elia and Piccolo, 2005). Thus, the parametric space is:

$$\Omega(\pi,\xi) = \{(\pi,\xi): 0 < \pi \leq 1;\ 0 \leq \xi \leq 1\}.$$

From a theoretical point of view, Iannario (2009c) proved that CUB models are fully identifiable for any $m > 3$. Moreover, the proposed mixture distribution is rather flexible and, depending on the parameters, it is able to assume very different shapes: symmetric or extremely skewed, rather flat or with definite mode, and this fact makes it a very useful tool for describing observed ordinal data.

As mentioned in the previous section, $(1-\pi)$ is a *measure of uncertainty* whereas $(1-\xi)$ may be interpreted as a *measure of performance*. Considering the whole random variable support, $(1-\pi)/m$ is a measure of the related *uncertainty share*.

The interpretation of ξ needs some caution because it depends on the initial coding of the responses (as a matter of fact, the graded scale may represent the strongest feeling/concern either by the highest value or by the lowest value). In particular, in several studies conducted in various fields, the parameter ξ has been related to the degree of perception, the strength of selectiveness/awareness, the measure of concern and the threshold of pain.

The parameter values help to locate CUB models in the parametric space defined by the unit square. This is a convenient way of giving an interpretation to results since it allows immediate comparisons among probability structures describing observed ratings. Thus, since $1-\pi$ quantifies the *propensity* of respondents to behave in accordance to a completely random choice, the more π is located to the right side of the unit square, the more respondents give definite answers (uncertainty is low). Similarly, since $1-\xi$ measures the strength of feeling of the subjects for a direct and positive evaluation of the object, the closer ξ is located to the border of the upper region of the unit square the less the item has been preferred.

Fitting to observed ordinal data usually improves when the subjects' *covariates* are introduced in order to relate both the feeling and the uncertainty to the respondents' features. Besides the presence of significant covariates helps the model interpretation and the discrimination among different sub-populations. The latter aim is accomplished by using dummy covariates (Iannario, 2008b) or by clustering methods (Corduas, 2008c,b). In addition, objects' covariates may be introduced (Piccolo and D'Elia, 2008) and thus, similarly to other contexts, CUB models may include *choices' covariates* and *chooser's covariates*: Agresti (2002).

In this regard, we should observe that the expected value of Y is given by:

$$\mathbb{E}(Y) = \pi(m-1)\left(\frac{1}{2}-\xi\right) + \frac{(m+1)}{2}.$$

Consequently, different parameter vectors $\boldsymbol{\theta} = (\pi,\xi)'$ may generate the same mean value. In such a context, it would not be therefore correct to introduce a link among expectation and covariates (as usually happens in GLM framework). In fact, CUB distributions can be rather different even if they have the same mean value. For this reason, we prefer a more general framework (advocated by King et al., 2000) where parameters describing the probability distribution are directly related to covariates.

Then, the general formulation of a CUB (p,q) model (with p covariates to explain uncertainty and q covariates to explain feeling) is expressed by the *stochastic component*:

$$Pr(Y=y \mid \boldsymbol{x}_i; \boldsymbol{w}_i) = \pi_i \binom{m-1}{y-1} \xi_i^{m-y}(1-\xi_i)^{y-1} + (1-\pi_i)\left(\frac{1}{m}\right), \quad y=1,2,\ldots,m,$$

and two *systematic components*:

$$\pi_i = \frac{1}{1+e^{-\boldsymbol{x}_i\boldsymbol{\beta}}}; \quad \xi_i = \frac{1}{1+e^{-\boldsymbol{w}_i\boldsymbol{\gamma}}}; \quad i=1,2,\ldots,n;$$

where $\boldsymbol{x}_i$ and $\boldsymbol{w}_i$ are the subjects' covariates for explaining π_i e ξ_i, respectively (Table 7.1).

Table 7.1 Notation of CUB (p,q) models, without and with covariates

Models	Covariates	Parameters	Parameter spaces
CUB $(0,0)$	No covariates	$\boldsymbol{\theta}=(\pi,\xi)'$	$(0,1] \times [0,1]$
CUB $(p,0)$	Only for π	$\boldsymbol{\theta}=(\boldsymbol{\beta}',\xi)'$	$\mathbb{R}^{p+1} \times [0,1]$
CUB $(0,q)$	Only for ξ	$\boldsymbol{\theta}=(\pi,\boldsymbol{\gamma}')'$	$(0,1] \times \mathbb{R}^{q+1}$
CUB (p,q)	For π and ξ	$\boldsymbol{\theta}=(\boldsymbol{\beta}',\boldsymbol{\gamma}')'$	$\mathbb{R}^{p+q+2}$

Notice that this formalization allows that the two sets of covariates may present some overlapping.

The nature of the probability distributions (Uniform and shifted Binomial) included in the mixture and the presence of Covariates justify the acronym CUB (the acronym MUB was used in some initial contributions).

With respect to the classical GLM approach (where proportional, adjacent or continuation ratio probabilities are introduced for ordinal data), CUB models offer a straightforward relationship between a probability statement for ordinal answers and subjects' covariates by means of a monotone function (logistic function, in most cases). Moreover, although latent variables are conceptually necessary in order to specify the nature of the mixture components, the inferential procedures are not based upon the knowledge (or estimation) of cut-points. As a consequence, when the CUB model turns out to be adequate in fitting data, it is usually more parsimonious with respect to models derived by the GLM approach.

CUB models have been further generalized for taking the possible effect of atypical situations into account. Sometimes, these are derived by *shelter choices*, which represent categories frequently selected by respondents in order to avoid more elaborate decisions.

Specifically, an *extended* CUB *model* is defined by:

$$p_y(\boldsymbol{\theta}) = \pi_1 \binom{m-1}{y-1} \xi^{m-y}(1-\xi)^{y-1} + \pi_2 \frac{1}{m} + (1-\pi_1-\pi_2) D_y^{(c)}, \quad y=1,2,\ldots,m,$$

where $\boldsymbol{\theta} = (\pi_1, \pi_2, \xi)'$ is the parameter vector characterizing the distribution of this new mixture random variable and $D_y^{(c)}$ is a degenerate random variable whose probability mass is concentrated at $y = c$, that is:

$$D_y^{(c)} = \begin{cases} 1, & \text{if } y = c; \\ 0, & \text{otherwise.} \end{cases}$$

We observe that extended CUB models are identifiable only for $m > 4$.

Of course, if $\pi_1 + \pi_2 = 1$ the extended CUB model collapses to the standard one. Instead, if $\pi_2 = 0$ we are just considering a mixture of a shifted Binomial distribution and a degenerate probability with mass at $(Y = c)$. Moreover, if $\pi_1 = \pi_2 = 0$ the extended model is able to account also for the (rare) situation where most of respondents' choices are concentrated at a single intermediate category.

A remarkable feature of the extended model is that parameter $\delta = 1 - \pi_1 - \pi_2$ measures the added relative contribution of the *shelter choice* at $y = c$ with respect to the standard version of the model. Since its significance may be tested via standard asymptotic inference, extended CUB models may check the effective relevance of the presence of a *shelter choice*. Furthermore, it should be noted that in some circumstances – if one avoids considering this component – parameter estimates are biased and inefficient, and fitting and predictions are not satisfactory.

Among others, this effect has been found in the evaluations of a data set collected among students attending courses at the University of Naples Federico II. The main objective of the survey was to measure several aspects of students' satisfaction with the teaching, lecture halls, time scheduling, services, etc. The survey was conducted using a questionnaire where the assessment of each item was based on the following 7 points scale: "extremely unsatisfied" ($=1$), "very unsatisfied" ($=2$), "unsatisfied" ($=3$), "indifferent" ($=4$), "satisfied" ($=5$), "very satisfied" ($=6$), "extremely satisfied" ($=7$). Thus, the assessment of a given item generates a rating Y with $m = 7$. In general, it has been observed that the distributions for most of the items under investigation present a very marked mode at $Y = 5$ (corresponding to the "satisfied" category).

Since respondents were a selected subset of enrolled students (those who regularly attend lectures are more likely to be satisfied with University life), a consistent part of them preferred to select the first positive judgement available on the proposed graded scale in order to avoid a more thoughtful assessment. In these cases, one should test the hypothesis $H_0 : \delta = 0$ in the extended CUB model with $c = 5$. In the examined data set, the parameter estimate $\hat{\delta} = 0.223$ (with a standard error of 0.004) confirms a substantial effect of the *shelter choice* with respect to the expected one. Moreover, the model fitting and the prediction of expected responses are improved.

A final remark concerns the possible presence of bimodal (multimodal) distributions which, at a first sight, may suggest adding further Binomial components to the mixture distribution in order to model the presence of various modes. In our opinion, adding random variables of the same family in order to explain the different behavior of respondents should be avoided since problems concerning model identifiability may arise. Instead, for this purpose, the introduction of subjects covariates should be seriously considered so that clustered responses might be taken into account. For instance, when people are asked to give a rate to a politician, the bimodal distributions of responses may be easily modelled if the ideological position (left/right) of the respondents are surveyed. In such a case, dichotomous or polytomous variables will be introduced as explanatory variables in the CUB model in order to explain the opposite expressed feeling.

7.4 Inferential issues and numerical procedures

Given a sample of observed ordinal data and covariates $(y_i, \boldsymbol{x}_i, \boldsymbol{w}_i)'$, for $i = 1, 2, \ldots, n$, the log-likelihood function for the parameter vector $\boldsymbol{\theta} = (\boldsymbol{\beta}', \boldsymbol{\gamma}')'$ in a general CUB (p, q) model is defined by:

$$\ell(\boldsymbol{\theta}) = \sum_{i=1}^{n} \log \left[\frac{1}{1 + e^{-\boldsymbol{x}_i \boldsymbol{\beta}}} \left\{ \binom{m-1}{y_i - 1} \frac{e^{(-\boldsymbol{w}_i \boldsymbol{\gamma})(y_i - 1)}}{(1 + e^{-\boldsymbol{w}_i \boldsymbol{\gamma}})^{m-1}} - \frac{1}{m} \right\} + \frac{1}{m} \right].$$

Inferential issues for the joint efficient estimation of the parameters are discussed in details by Piccolo (2006) who derived the EM algorithm for maximum likelihood (ML) estimation. The procedure is effective but convergence to maximum can be rather slow; then, several proposals for improving preliminary parameter estimates have been suggested in order to improve the rate of convergence (Iannario, 2009a). In this regard, moment estimators provide useful initial values but some problems arise for models with covariates. These aspects are currently under investigation.

In the following section, a brief illustration of the EM estimation algorithm is presented with special reference to the extended model without covariates.

7.4.1 The EM algorithm

Let $\boldsymbol{y} = (y_1, y_2, \ldots, y_n)'$ be the sample of ordinal data generated by a survey where n respondents are asked to choose an integer in the support $\{1, 2, \ldots, m\}$, for a given $m > 4$. We suppose the location $c \in \{1, 2, \ldots, m\}$ of the *shelter choice* is known.

For the extended CUB model, the log-likelihood function $\ell(\boldsymbol{\theta})$ for the sample $\boldsymbol{y}$, with $\boldsymbol{\theta} = (\pi_1, \pi_2, \xi)'$, is

$$\ell(\boldsymbol{\theta}) = \sum_{i=1}^{n} \log\left[Pr(Y=y_i \mid \boldsymbol{\theta})\right]$$
$$= \sum_{i=1}^{n} \log\left[\pi_1 b_{y_i}(\xi) + \pi_2 U_{y_i}(m) + (1-\pi_1-\pi_2) D_{y_i}^{(c)}\right],$$

where the components of the mixture are specified, for $i = 1, 2, \ldots, n$, by:

$$p_g(y_i; \boldsymbol{\theta}_g) = \begin{cases} b_{y_i}(\xi) = \binom{m-1}{y_i-1} \xi^{m-y_i}(1-\xi)^{y_i-1}, & g=1; \\ U_{y_i}(m) = \frac{1}{m}, & g=2; \\ D_{y_i}^{(c)}, & g=3. \end{cases}$$

We introduce the unobservable vector $\mathbf{z} = (\mathbf{z}_1, \mathbf{z}_2, \ldots, \mathbf{z}_n)'$ where each $\mathbf{z}_i = (z_{1i}, z_{2i}, z_{3i})'$ is a three-dimensional vector such that, for $g = 1, 2, 3$:

$$z_{gi} = \begin{cases} 1, & \text{if the } i\text{-th subject belongs to the } g \text{ group;} \\ 0, & \text{otherwise.} \end{cases}$$

Simplifying the notation, we let:

$$\pi_g = \begin{cases} \pi_1, & g=1; \\ \pi_2, & g=2; \\ 1-\pi_1-\pi_2, & g=3. \end{cases} \qquad \boldsymbol{\theta}_g = \begin{cases} \boldsymbol{\theta}_1 = (\pi_1, \xi)', & g=1; \\ \boldsymbol{\theta}_2 = \pi_2, & g=2; \\ \boldsymbol{\theta}_3 = 1-\pi_1-\pi_2, & g=3. \end{cases}$$

Then, the likelihood function of the complete-data vector $(\mathbf{y}', \mathbf{z}')'$ is given by:

$$L_c(\boldsymbol{\theta}) = \prod_{g=1}^{3} \prod_{i=1}^{n} [\pi_g p_g(y_i; \boldsymbol{\theta}_g)]^{z_{gi}},$$

and the complete-data log-likelihood function is:

$$\ell_c(\boldsymbol{\theta}) = \sum_{g=1}^{3} \sum_{i=1}^{n} z_{gi} \left[\log(\pi_g) + \log(p_g(y_i; \boldsymbol{\theta}_g))\right].$$

The $(k+1)$-th iteration of the EM algorithm consists of the following steps:

- *E-step*:

The conditional expectation of the indicator random variable Z_{gi}, given the observed sample $\mathbf{y}$, is:

$$\mathbb{E}\left(Z_{gi} \mid \mathbf{y}, \boldsymbol{\theta}^{(k)}\right) = Pr\left(Z_{gi} = 1 \mid \mathbf{y}, \boldsymbol{\theta}^{(k)}\right) = \frac{\pi_g^{(k)} p_g\left(\mathbf{y}; \boldsymbol{\theta}_g^{(k)}\right)}{\sum_{j=1}^{3} \pi_j^{(k)} p_j\left(\mathbf{y}; \boldsymbol{\theta}_j^{(k)}\right)} = \tau_g(y_i; \boldsymbol{\theta}^{(k)}),$$

for any $g = 1, 2, 3$. This quantity is the posterior probability that the i-th subject of the sample with the observed y_i belongs to the g-th component of the mixture.

Then, the expected log-likelihood of complete-data vector is obtained as:

$$\mathbb{E}(\ell_c(\boldsymbol{\theta})) = \sum_{g=1}^{3} \sum_{i=1}^{n} \tau_g(y_i; \boldsymbol{\theta}^{(k)}) \left[\log\left(\pi_g^{(k)}\right) + \log\left(p_g\left(y_i; \boldsymbol{\theta}_g^{(k)}\right)\right) \right].$$

- **M-step**:

At the $(k+1)$-th iteration, the function $Q(\boldsymbol{\theta}^{(k)}) = \mathbb{E}(\ell_c(\boldsymbol{\theta}))$ has to be maximized with respect to the parameters (π_1, π_2) and ξ. If the parameters of the components are specified, this quantity may be expressed as follows:

$$Q(\boldsymbol{\theta}^{(k)}) = \sum_{i=1}^{n} \left[\tau_1\left(y_i; \boldsymbol{\theta}_1^{(k)}\right) \log\left(\pi_1^{(k)}\right) + \tau_2\left(y_i; \boldsymbol{\theta}_2^{(k)}\right) \log\left(\pi_2^{(k)}\right) + \tau_3\left(y_i; \boldsymbol{\theta}_3^{(k)}\right) \log\left(\pi_3^{(k)}\right) \right]$$

$$+ \sum_{i=1}^{n} \sum_{g=1}^{3} \left[\tau_g\left(y_i; \boldsymbol{\theta}_g^{(k)}\right) \log\left(p_g\left(y_i; \boldsymbol{\theta}_g^{(k)}\right)\right) \right]$$

$$= S_1 \log\left(\pi_1^{(k)}\right) + S_2 \log\left(\pi_2^{(k)}\right) + (n - S_1 - S_2) \log\left(1 - \pi_1^{(k)} - \pi_2^{(k)}\right) + Q^*$$

where Q^* is independent from $\pi_g^{(k)}$ parameters, and

$$S_g = \sum_{i=1}^{n} \tau_g(y_i; \boldsymbol{\theta}^{(k)}), \quad g = 1, 2; \qquad S_3 = n - S_1 - S_2.$$

Then, by solving the system: $\frac{\partial Q(\boldsymbol{\theta}^{(k)})}{\partial \pi_g} = 0$, for $g = 1, 2$, we get:

$$\pi_1^{(k+1)} = \frac{S_1}{n} = \frac{1}{n} \sum_{i=1}^{n} \tau_1(y_i; \boldsymbol{\theta}^{(k)}); \qquad \pi_2^{(k+1)} = \frac{S_2}{n} = \frac{1}{n} \sum_{i=1}^{n} \tau_2(y_i; \boldsymbol{\theta}^{(k)}).$$

Instead, the estimate of ξ, for a given k, is obtained from:

$$\sum_{i=1}^{n} \tau_1\left(y_i; \boldsymbol{\theta}_1^{(k)}\right) \frac{\partial \log(p_1(y_i; \xi))}{\partial \xi} = 0.$$

A simple algebra produces the solution:

$$\xi^{(k+1)} = \frac{m - \overline{Y}_n(p)}{m - 1}; \qquad \overline{Y}_n(p) = \frac{\sum_{i=1}^{n} y_i \tau_1\left(y_i; \boldsymbol{\theta}_1^{(k)}\right)}{\sum_{i=1}^{n} \tau_1\left(y_i; \boldsymbol{\theta}_1^{(k)}\right)}.$$

Here, $\overline{Y}_n(p)$ is the average of the observed sampled values weighted with the posterior probability that y_i is a realization of the first component of the mixture (that is a shifted Binomial distribution).

Then, E- and M- steps are repeated with new parameters $\left(\pi_1^{(k+1)}, \pi_2^{(k+1)}, \xi^{(k+1)}\right)'$ until a convergence criterion is satisfied. For instance, this could be given by:
$|\ell(\boldsymbol{\theta}^{(k+1)}) - \ell(\boldsymbol{\theta}^{(k)})| < \varepsilon$, for a small $\varepsilon > 0$.

Notice that, as far as ML estimation is concerned, sample data $(y_1, y_2, \ldots, y_n)'$ is equivalently represented by the vector of absolute frequencies $(n_1, n_2, \ldots, n_m)'$. For computational efficiency, it is therefore convenient to use in previous steps the log-likelihood function for grouped data. To this end, we will compute:

$$S_g = \sum_{y=1}^{m} n_y \, \tau_g(y; \boldsymbol{\theta}^{(k)}), \quad g = 1, 2.$$

The step-by-step formulation of the EM algorithm may be easily programmed in formal languages (such as *GAUSS©*, *Matlab©* or *R*).

Maximum likelihood inference has been developed by using standard approaches (Piccolo, 2006). Specifically, the asymptotic variance-covariance matrix $\boldsymbol{V}(\boldsymbol{\theta})$ of ML estimators $\hat{\boldsymbol{\theta}}$ of the parameter $\boldsymbol{\theta}$ of CUB model is based on the *observed information matrix* $\mathscr{I}(\boldsymbol{\theta})$, that is the negative of the Hessian computed at $\boldsymbol{\theta} = \hat{\boldsymbol{\theta}}$; it shares the same asymptotic properties of the expected information matrix (as argued by Pawitan (2001), 244–247 among others).

Then, the asymptotic variance-covariance matrix $\boldsymbol{V}(\boldsymbol{\theta})$ of the ML estimators of $\boldsymbol{\theta}$, computed at $\boldsymbol{\theta} = \hat{\boldsymbol{\theta}} = (\hat{\pi}, \hat{\xi})'$, is obtained as:

$$\boldsymbol{V}(\boldsymbol{\theta}) = \left[\mathscr{I}(\hat{\boldsymbol{\theta}})\right]^{-1} = -\begin{pmatrix} \frac{\partial^2 \ell(\boldsymbol{\theta})}{\partial \pi^2} & \frac{\partial^2 \ell(\boldsymbol{\theta})}{\partial \pi \partial \xi} \\ \frac{\partial^2 \ell(\boldsymbol{\theta})}{\partial \pi \partial \xi} & \frac{\partial^2 \ell(\boldsymbol{\theta})}{\partial \xi^2} \end{pmatrix}^{-1}_{(\boldsymbol{\theta}=\hat{\boldsymbol{\theta}})}.$$

The computational details for implementing these results are discussed by Piccolo (2006) and a related software in *R* is currently available for estimation and inference about CUB models with (or without) covariates (Iannario and Piccolo 2009).

7.4.2 Fitting measures

The adequacy of models may be checked by means of several measures (significance of parameters, sensible increase in log-likelihood, and so on). However, the sample size of evaluation data sets is generally large and thus, in order to verify how estimated CUB models fit empirical data, we prefer to introduce a descriptive measure for models without covariates and refer to likelihood-based indexes for more general comparisons.

Specifically, from a descriptive point of view, we consider the normed *dissimilarity index* Diss $\in [0, 1]$ defined by:

$$\text{Diss} = \frac{1}{2} \sum_{y=1}^{m} \left| Pr\left(Y = y \mid \hat{\boldsymbol{\theta}}\right) - \frac{n_y}{n} \right|.$$

This index has an appealing interpretation since it measures the proportion of subjects that should modify their choices in order to reach a perfect fit between observed and theoretical distributions (Leti, 1979; Simonoff, 2003). Unfortunately, it cannot be immediately extended to the case of CUB models with covariates.

For this aim, using an obvious notation, log-likelihoods of CUB models can be compared as follows:

Comparisons	Deviances difference	Degrees of freedom
CUB $(p,0)$ versus CUB $(0,0)$	$2(\ell_{10} - \ell_{00})$	p
CUB $(0,q)$ versus CUB $(0,0)$	$2(\ell_{01} - \ell_{00})$	q
CUB (p,q) versus CUB $(0,0)$	$2(\ell_{11} - \ell_{00})$	$p+q$

The difference between deviances should be compared with the quantile of the χ^2 distribution with degrees of freedom as reported in the table above.

In this regard, the log-likelihood for the *saturated* CUB model can provide a useful benchmark:

$$\ell_{\text{sat}} = -n \log(n) + \sum_{y=1}^{m} n_y \log(n_y).$$

The fitting measure may be obtained by defining a pseudo-R^2, that is named ICON (=*I*nformation CONtent), which compares the log-likelihood of the estimated model with the log-likelihood of a discrete Uniform random variable fitted to data (this is in fact the uninformative model). Thus, the ICON index is:

$$\text{ICON} = 1 + \frac{\ell(\hat{\boldsymbol{\theta}})/n}{\log(m)}.$$

It measures the improvement achieved by a CUB model, without or with covariates, with respect to a completely uninformative distribution (such as the Uniform distribution). In other words, this index is related to the displacement of the log-likelihood of the estimated model with respect to an extreme situation.

7.5 Fields of application

In opinion surveys people are often requested to arrange a list of m items in order of preferences or, alternatively, they are asked to express judgements or evaluations using a given m-point ordinal scale. In this respect, we need to distinguish clearly between two situations: the *rating* where the subject's answer is a single score for each item, and the *ranking* where the answer is a permutation of the first m integers, that is a vector of numbers specifying sequentially the degree of preferences for the m objects.

For the correct understanding of the usage of CUB models, it is important to underline that our approach suggests a mixture distribution useful for modelling the random variable generated by the assessment of a single item (*rating*) or by the positions of a single object in the ordering (*ranking*). However, notice that while in the first case the CUB model is applied to study the univariate response of a group of subjects, in the second case the model is used for the marginal analysis of the discrete multivariate random variable generated by the observed preferences for m objects. In the latter case, it is evident that adopting this strategy in turn for all the m marginal distributions of the ranks leads to non independent random variables.

In previous studies, various applications of the proposed approach have been elaborated in order to fit and interpret univariate *rating data*, especially in relation to evaluations of attributes of goods and services (Corduas, 2008c) and other fields of analysis such as social analysis (Iannario, 2007, 2008a), medicine (D'Elia, 2008), sensometric studies (Piccolo and D'Elia, 2008) and linguistics (Balirano and Corduas, 2008). In such contexts, the paradigm based on modelling the feeling and uncertainty components has turned out to be very useful for its interpretative content.

A further kind of application stems from *categorical data* that are qualitative in their nature although they are actually measured by means of a quantitative scale. In these cases, a genuine ordinal approach proves to be more fruitful for the interpretation and assessment of original data. This approach has been pursued in order to investigate how final grades achieved by university students are related to gender and time spent to complete the university program of studies. As a matter of fact, although the final grade is expressed on a quantitative scale it should be regarded as a qualitative assessment of the examining committee about candidates. This case study has confirmed the better performance of qualitative models with respect to standard quantitative models in relation to the tails of the distribution (given the robustness property of ordinal values) and to the prediction of extreme data.

Finally, the transformation of subjective survival probability (expressed by a percentage on $[0, 100]$ scale) into an ordinal score described by a standard 7-point scale has provided another interesting data base for further modelling. This is, in fact, a typical case where the numerical value, that a subject gives in reply to a question, is clearly generated by a qualitative consideration about the perception he/she has of "high" or "low" probability. Again, CUB models has proved to be effective.

7.6 Further developments: a clustering approach

In order to compare rating distributions related to a number of items or to different groups of respondents, a clustering procedure for ordinal data based on estimated CUB models has been introduced by Corduas (2008d).

The search for a special approach is motivated by the risk of misleading interpretations of data arising from the representation of CUB models in the parameter space, because in such a situation the user tends to assess the closeness of two (estimated)

CUB distributions in terms of the Euclidean distance between the corresponding estimated parameters. As a matter of fact, the variability of π and ξ estimators are different and, in addition, the role of CUB coefficients in determining the shape of the estimated distribution is very dissimilar (Piccolo, 2003).

For this reason, Kullback-Liebler (KL) divergence can be used for testing similarities among distributions. Consider two discrete populations each characterized by a probability distribution function having the same functional form $p(y, \boldsymbol{\theta}_i)$ with unspecified parameters $\boldsymbol{\theta}_i$, $i = 1, 2$. Also assume that $p(y, \boldsymbol{\theta}_i) > 0$, $\forall y$. Suppose that two samples of n_1 and n_2 observations have been randomly drawn from each population, respectively. In order to test the hypothesis $H_0 : \boldsymbol{\theta}_1 = \boldsymbol{\theta}_2$, the KL divergence statistic is defined by:

$$\widehat{J} = \frac{n_1 n_2}{n_1 + n_2} \left[\sum_y \left(p(y; \boldsymbol{\theta}_1) - p(y; \boldsymbol{\theta}_2) \right) \log \frac{p(y; \boldsymbol{\theta}_1)}{p(y; \boldsymbol{\theta}_2)} \right]_{(\boldsymbol{\theta}_1 = \hat{\boldsymbol{\theta}}_1, \boldsymbol{\theta}_2 = \hat{\boldsymbol{\theta}}_2)}$$

where the parameters $\boldsymbol{\theta}_1$ and $\boldsymbol{\theta}_2$ have been replaced by the ML estimators. Under the null hypothesis, it can be shown that $\widehat{J}$ is asymptotically distributed as a $\chi^2_{(g)}$ random variable (Kullback, 1959), being g the common dimension of the parameter vector; in the special case under investigation $g = 2$.

The strategy for grouping a set of CUB models combines hypotheses testing with a clustering algorithm. Firstly, $\widehat{J}$ for each couple of models is evaluated. Secondly, a binary matrix is built by setting the (i, j)th entry equal to 0, when the hypothesis of homogeneity of the i-th and j-th models is rejected, and 1 otherwise. Finally, by means of convenient algorithms (such as BEA: McCormick et al., 1972; Arabie and Hubert, 1990), this matrix is rearranged into an approximate block diagonal form. A clearly defined (unit) triangle immediately under the diagonal will indicate a cluster of items for which the judgements expressed by respondents (and summarized by CUB distributions) are similar. The presence of any zero value in such a triangle indicates that the cluster may be elongated or constituted by other well separated small clusters.

The proposed technique is able to discriminate the different patterns of the rating distributions with respect to skewness, kurtosis, mode and it is very effective and selective as has been proved by various empirical studies (for instance, see Corduas, 2008b,c,d for a study concerning university students' opinions about teaching quality) .

7.7 Case study

In the years 2002–2004 and 2007–2008, the University of Naples Federico II carried out an extensive survey of students' opinions concerning the Orientation services which operated in the 13 Faculties. In this section, the study will focus on the data sets gathered in the last 2-year period.

7 A class of statistical models for evaluating services and performances

A questionnaire was submitted to a sample of users and each student was asked to give a score for expressing his/her satisfaction with various aspects of the Orientation service. Eight items have been investigated: staff willingness (=WILL) and competence (=COMPE), clearness of information (=INFO), suitable opening hours (=TIME), adequate equipment and structure (=STRU), advertisement of the service (=ADVE), usefulness of information for decisions (=DECI), and a final overall evaluation (=GLOBA). Judgments were expressed using the ordinal scale ranging from 1 (="completely unsatisfied") to 7 (="completely satisfied").

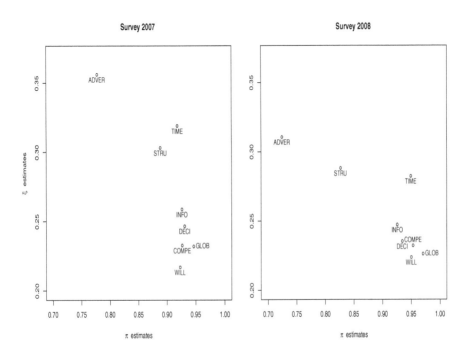

Fig. 7.1 CUB models of students' satisfaction with University Orientation services.

In Fig. 7.1, the estimated parameters of CUB models built for the eight items for 2007 and 2008 surveys are plotted in the parameter space. The results refer to 3,511 and 4,042 validated questionnaires for the first and second survey, respectively.

Respondents show a different attitude towards the activities performed by the staff and the aspects related to office organization and equipment since the former type of items systematically receive higher evaluations than the others.

Moreover, comparing the results from the first and the second survey, the expressed satisfaction with items concerning office organization and equipment seems to improve. In Fig. 7.1, the corresponding estimated CUB models, in fact, moves from the top part of the graph to the lower one. Noticeably, the opinion for the lack of adequate advertisements of the service is very critical. Finally, items related to staff evaluations receive more resolute assessment in the second year; the estimated value of $(1-\xi)$ for this type of items is higher than for any other.

Furthermore, the plot suggests that at least two latent variables may govern the responses since the models appear to be grouped in two separate clusters.

The merit of the previous examples is that CUB models are able to summarize and visualize rating distributions originating from thousands of opinions given in different periods of time.

Afterwards, attention is concentrated on 2007 data sets which have been partly examined by Iannario and Piccolo (2008). We conjecture that the CUB model related to students' satisfaction with the office opening hours (=TIME) may improve by introducing significant covariates. Then, available covariates have been added to the CUB model by a stepwise strategy. Specifically, the covariate that mostly improves the log-likelihood function, compared to the others, has been preferred. The resulting parameter estimates (in parentheses their standard errors) and the corresponding log-likelihood values are presented in Table 7.2. Comparison of deviances (not reported here) confirms that the fitted models are all significant and better than the nested ones.

Table 7.2 CUB (p,q) models of students' evaluation for *opening hours*

Models	$\hat{\pi}$	$\hat{\xi}(w)$	log-likelihood
▶ CUB $(0,0)$	0.918 (0.011)	$\hat{\xi} = 0.319\ (0.004)$	$\ell_{00} = -5714.8$
▶ CUB $(0,1)$ log(Age)	0.920 (0.011)	$\hat{\gamma}_0 = 1.464\ (0.347)$ $\hat{\gamma}_1 = -0.722\ (0.113)$	$\ell_{01} = -5693.6$
▶ CUB $(0,2)$ log(Age) Gender	0.921 (0.010)	$\hat{\gamma}_0 = 1.505\ (0.348)$ $\hat{\gamma}_1 = -0.756\ (0.114)$ $\hat{\gamma}_2 = 0.116\ (0.034)$	$\ell_{02} = -5687.6$
▶ CUB $(0,3)$ log(Age) Gender Change	0.921 (0.010)	$\hat{\gamma}_0 = 1.601\ (0.349)$ $\hat{\gamma}_1 = -0.793\ (0.114)$ $\hat{\gamma}_2 = 0.114\ (0.034)$ $\hat{\gamma}_3 = 0.190\ (0.054)$	$\ell_{03} = -5681.6$
▶ CUB $(0,4)$ log(Age) Gender Change Full-time (FT)	0.922 (0.010)	$\hat{\gamma}_0 = 1.879\ (0.375)$ $\hat{\gamma}_1 = -0.866\ (0.120)$ $\hat{\gamma}_2 = 0.116\ (0.034)$ $\hat{\gamma}_3 = 0.182\ (0.054)$ $\hat{\gamma}_4 = -0.078\ (0.038)$	$\ell_{04} = -5679.4$

We denote the covariates for the i-th subject as:

$$\boldsymbol{w}_i = (\log(\text{Age}_i), \text{Gender}_i, \text{Change}_i, FT_i)'.$$

Then, given $m = 7$, the best CUB $(0,4)$ model implies the following probability distributions for the expressed evaluations:

$$Pr(Y = y \mid \boldsymbol{w}_i) = 0.011 + 0.922 \binom{6}{y-1}(1-\xi_i)^{y-1}\xi_i^{7-y}, \ y = 1, 2, \ldots, 7,$$

where the parameters $\xi_i = \xi_i \mid \boldsymbol{w}_i$, $i = 1, 2, \ldots, n$, are specified by:

$$\frac{1}{1+\exp\{-1.879+0.866\log(\text{Age}_i)-0.116\,\text{Gender}_i-0.182\,\text{Change}_i+0.078\,FT_i\}}$$

Since $(1-\xi)$ is a measure of satisfaction, the estimated model shows that evaluation increases with *Age* and for full-time students $(FT = 1)$ whereas women (Gender $= 1$) and students who change their original enrollment and move from one Faculty to another (Change $= 1$) lower their preferences, and thus they are more critical about Opening Hours.

Table 7.3 Comparison of different students' profiles and corresponding parameters

Profiles	Age	Gender	Change	Full-time	$\boldsymbol{w}_i$	$\xi_i \mid \boldsymbol{w}_i$	$Pr(Y \geq 5)$
A	20	Woman	No	Yes	$(20,1,0,1)'$	0.337	0.654
B	40	Woman	No	Yes	$(40,1,0,1)'$	0.218	0.843
C	20	Man	Yes	Yes	$(20,0,1,1)'$	0.352	0.627
D	40	Man	Yes	Yes	$(40,0,1,1)'$	0.229	0.828
E	20	Woman	Yes	Yes	$(40,1,1,1)'$	0.379	0.576
F	40	Woman	Yes	Yes	$(40,1,1,1)'$	0.251	0.798

The model allows immediate comparison of different profiles; some of them are proposed in Table 7.3. Notice that the implied coefficient $\pi = 0.922$ is constant for all profiles since there are no significant covariates for the uncertainty component in the best estimated model.

It is evident from Fig. 7.2 how the age of the student is the relevant covariate forcing rating distribution into higher values. The last column in Table 7.3 shows that the probability of a positive evaluation mostly changes with age. Marginal changes in the distribution shape are determined by job position and by changing the original university enrollment to enter a new Faculty. Because of the large sample size, these covariates are significant although they achieve a modest impact.

The examination of expected evaluation for given profiles of respondents allows further considerations about the use of CUB models in empirical studies. As far as ordinal variables are concerned, expected values should only be considered for comparative purposes rather than being used as an index which is meaningful in itself.

In the present work, ordinal variables are intended as a monotone transformation of a latent variable Y^* then the study of the expected value of the random variable Y is worthy of interest whenever it is referred to groups of respondents with the same profile.

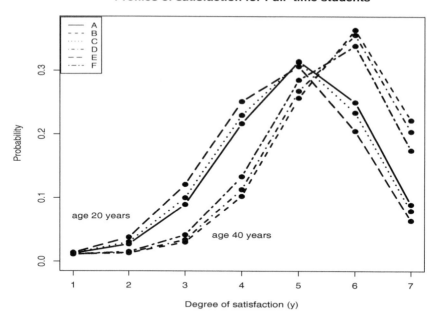

Fig. 7.2 CUB models of students' satisfaction with university orientation services.

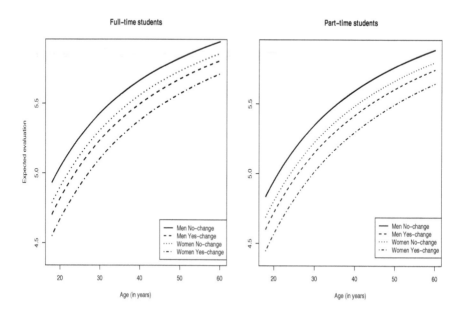

Fig. 7.3 Expected student satisfaction with opening hours for given covariates.

Figure 7.3 exemplifies this approach. In particular, the expected satisfaction is shown for the varying age of the students and their significant covariates. The plots confirm that satisfaction improves with age in a systematic way, a small increase may be observed for full-time students and those that did not change Faculty and a more severe judgement is formulated by women compared to men.

7.8 Concluding remarks

Although CUB models only describe univariate distributions of judgements, their use seems to be effective for investigating sound relationships among ordinal responses and covariates and, in addition, for enhancing unobserved traits in the data. In particular, the role of covariates is made manifest in the model and result in a useful device for the analysis of profiles.

Some unexplored issues that deserve further research are worth mentioning:

- Evaluation data and performances measures are collected in stratified subgroups both for economic reasons and research needs. Then, the introduction of multilevel CUB models is a relevant issue for further developments in this area.
- It is well-known that the range of multivariate distributions implied by the given marginal CUB models is limited. Thus, the efforts for generalizing the approach to a multivariate framework should help to retain the effectiveness of this parametrization and improve current interpretations.
- Fitting measures should be examined closely in order to exploit information carried by likelihood functions for sampled data.
- Differences and areas of complementary usage with other well-established approaches, such as Item Response Theory, are currently under investigation.
- Since large data sets with a great quantity of information about subjects are commonly available from surveys, further studies are needed in order to improve the criteria for the selection of significant covariates and the preliminary parameter estimation by considering both numerical algorithms and data mining procedures.

Chapter 8
Choices and conjoint analysis: critical aspects and recent developments

Rossella Berni and Riccardo Rivello

8.1 Introduction

In the literature, a large number of researchers and practitioners are dealing with preference measurements which are considered as one of the most general methods in order to study and improve the consumer's behaviour intended as the consumer's decision about improving his/her utility in changing a service or a product. Nevertheless, a wide range of preference measurements' methods is defined according to the specific aim of the research, or of the application, and the basic theoretical elements involved therein.

In particular, the preference theory must be evaluated according to the nature and definition of preference, namely revealed or stated preferences and, in case of stated preferences, we may distinguish between Contingent Valuation (CV), Conjoint Analysis (CA) and Choice Modelling (CM), Hanley et al. (2001), Netzer et al. (2008). Nevertheless, by considering CA and CM, since the fundamental elements of distinctions are positively overlapped or interchanged, the classification is not so clearly definable; this can be observed when these methods are generally defined as multi-attribute methods.

However, the preference measurements about a product or a service are usually related to a new product/service and the main distinction between CA and CM is the monetary evaluation, namely the Willingness to Pay (WTP), which is the quantitative expression of the respondents about their willingness to accept a change in the product/service concerned or in a single attribute.

Rossella Berni
Department of Statistics "G. Parenti", University of Florence, Viale Morgagni 59,
50134 Florence, Italy,
e-mail: berni@ds.unifi.it

Riccardo Rivello
Department of Statistics "G. Parenti", University of Florence, Viale Morgagni 59,
50134 Florence, Italy,
e-mail: rivello@ds.unifi.it

Furthermore, even though some steps and methods of these two techniques could be viewed as very similar, e.g. the experimental design, the basic elements of the experiments theory are defined and applied in both contexts taking into account, at the same time, that there exists many theoretical differentiations. Thus, the related statistical models were separately developed in the last decades, McFadden (1974), McFadden and Train (2000), Lenk et al. (1996), Greene and Hensher (2003); but, the recent developments in this field were mainly directed at improving common features, such as the heterogeneity of respondents and the complexity of the alternatives (profiles).

In this chapter, we focus on stated preferences (SP) and, namely, on CA and CM, carrying out a brief and critical review in order to clarify the distinctions, as well as to point out the common issues. In addition, we deal with the possibility of reaching the best profile in CA through the theory of statistical methods in the engineering field by considering the current situation and the user's preferences. Our proposal is discussed by showing an empirical example. In this context, we point out the presence in the literature of similar attempts, where the common issue is related to the statistical method applied, the Response Surface Methodology (RSM), Danaher (1997), Jiao et al. (2007), or to the general aim of creating a link between the needs of the manufacturer (design product/service stage) and the consumer/user's preferences, Michalek et al. (2005), Du et al. (2006).

This chapter is organized as follows: a literature review on CA and CM is presented in the second section, by pointing out the methods and recent developments related to CE and CA, respectively; in the third section, our proposal of applying RSM in a CA context is discussed in detail. Section four presents the data and results about the empirical example, while the concluding remarks are outlined in the final section.

8.2 Literature review

Many developments and improvements in consumer/user's preferences by considering the experimental design and the statistical modelling were achieved in the last two decades. Nevertheless, we mainly pay attention to the period 2000–2008, when methods and related applications gave an in-depth consideration to specific issues. Undoubtedly, a further and clear distinction must be made when we refer to preference measurements or, more in general, to the preference theory. Hence, we deal with Stated Preferences (SP), where we define as SP the preference of a respondent related to a hypothetical scenario shown as an alternative in a choice-set (CM) or presented as one of the suggested profiles (CA). However, in the literature, some recent developments are also reported in the Revealed Preference case, which is defined as the preference of the respondent about a real situation, such as in Scarpa et al. (2003).

Another more subtle differentiation is when we refer to CV, CA, CM. Contingent Valuation (CV) is defined as a method in which the respondent is asked to give his/her preference on a product by considering only its total price (mono-attribute method); on the other hand, when we refer to CM and CA, the respondent is asked to express his/her preference or choice about a product or a service by also evaluating the monetary impact of several attributes, and, therefore, the Willingness to Pay (WTP) may be estimated for each single aspect (multi-attribute valuation methods - MAVs).

In our context, we mainly consider the two CA and CM methods, by pointing out the further distinction within CM between Contingent Ranking (CR) and Choice Experiments (CE). The CE situation is related to a set of alternatives, called choice-set, which is selected from an experimental design; the respondent is asked to give his/her preference within each choice-set. The CR situation is applied when the respondent is asked not to give his/her preference, but he/she must rank or order the alternatives of the choice-set (obviously, in this case, each choice-set is comprised of more than two alternatives). The further distinction between the three types of response variables is a straightforward matter. In CA, rating (metric scale) and ranking (ordinal scale) are the preferred response variables, owing to the different framework of profiles. In the CM situation, choice (binary or not) and ranking are surely the conditioning response variables. Our expression "conditioning" means the corresponding statistical models involved in the analysis. Undoubtedly, CE is the preferred method in the literature and, consequently, the related theory has been largely developed in the last few years, by considering the experimental design with its optimality criteria and statistical models and first of all the class of Random Utility Models (RUM) and its variations, see Train (1998), McFadden and Train (2000), Boxall and Adamowicz (2002), Hynes et al. (2008), Wen and Koppelman (2001).

It is not irrelevant to point out that when a methodology is comprised of several theoretical steps, as CM and CA, these elements (mainly experimental designs and statistical models) are closely connected (Yu et al. (2009), Toubia and Hauser (2007)), and the properties of one design affect the corresponding model. When these properties do not exist in the design, this must be taken into account in the model. This is the case of an improvement in the design optimality specifically defined for a Mixed Multinomial Logit (MMNL), Sandor and Wedel (2002); on the other hand, when considering the respondents' heterogeneity, a specific design matrix for each respondent is planned (Sandor and Wedel (2005)), by including the heterogeneity evaluation directly in the design step instead of the model step.

However, as was said hereinabove, a different evolution has characterized the experimental CA and CM designs, even though some features are in common, such as specific methods, algorithms and models, in order to select alternatives, by considering the planning step (De Bruyn et al. (2008) or Toubia et al. (2007)) or the analysis of collected data, such as in Netzer and Srinivasan (2007), where a dynamic evaluation of the questionnaire through an Adaptive Self Explication method is performed in a Multi-Attribute context.

In Table 8.1, differentiations are summarized between CA and CM, namely CE and CR, by considering these first issues and the time developments.

As is shown in the summary (Table 8.1), where some specific features such as status-quo are not yet included, the preference theory is more articulated when

Table 8.1 MAV methods- A summary of recent developments

Steps	Conjoint Analysis-CA	Choice Experiments-CE	Contingent Ranking-CR
Preference	profile	alternative-choice	alternative-order
Dep.var.	rate; rank	choice	rank
Exp. design	factorial; frac. factorial	D-optimal; Local Bay. optimal; optimality ad-hoc	D-optimal
Stat. models	linear model; Hyerarchical Bayesian	Random Utility Model (RUM): Nested Logit (NL), Generalized NL	RUM: Rank Ordered Logit-Asc, Kernel logit
	finite-mixture model	RUM: Mixed-MNL; Latent Class Model (LCM)	RUM: Rank Ordered Logit-LC

considering all the steps within the three methods. Furthermore, having previously outlined the differentiations related to the type of preference and to the dependent variable, we may now observe that the experimental design step could be varied within these methods. Undoubtedly, CA is an easier task at this point: the theory and applications in the literature present above all developments and studies about the complexity and selection of profiles in the model step, (De Bruyn et al. (2008), Netzer and Srinivasan (2007)), i.e. some problems of complexity, such as preference uncertainty and conflicts solved through the evaluation of judgement time and response error in a rating task (Fischer et al. (2000)). The design of experiments is involved in order to create an orthogonal design (sometimes optimal) where all the created profiles, according to the set of attributes considered, are eventually reduced by applying a fractional factorial design of high Resolution.

The complexity of statistical models developed in the recent years, like finite-mixture models and hierarchical Bayes models, such as in Gilbride and Allenby (2004) and Lenk et al. (1996), in order to take account of the respondents' heterogeneity or the complexity of alternatives, or in Bradlow et al. (2004) for imputing missing levels of profiles, has not yet received in the literature an adequate response when considering the properties of the experimental design. Instead, a different situation is presented in the CM sector, namely in the CE method, where optimality criteria, above all D-optimality, ad-hoc algorithms and specified information matrices for the experimental design involved were entirely defined in 1990s (Zwerina et al. (1996)). Recent developments are related to the construction of optimal or near optimal designs with two-level attributes for binary choices in the presence of the first order interactions, Street and Burgess (2004), or when optimal designs are defined with mixed-level attributes, Burgess and Street (2005). Furthermore, a new optimum criterium is suggested, the M-optimality (Toubia and Hauser (2007)), where attempts in order to focus the planning by considering the manager's need were introduced; in fact, M-optimality means optimality manager. Note that a common feature of recent years is to create a link among designs and models together with the need of a guiding thread between manufacturers and consumers. In addition, it is not so irrelevant to quote the paper of Sandor and Wedel (2002) which reflects the strict connection between experimental designs and statistical models, because they suggest an experimental design with ad-hoc properties for a Mixed Multinomial Logit. This model, belonging to the class of Random Utility models, is certainly the most widely applied and developed model in recent years for the CE situation. Its success is easily explained when considering the theoretical results of McFadden and Train (2000), Train (1998) and the possibility, by adding additional random parameters, to study respondents' heterogeneity and the correlation structures due to repeated choices. The last developments of this model include its relationship with the latent class model, in order to create a finite number of respondent groups (Greene and Hensher (2003), Hynes et al. (2008), Boxall and Adamowicz (2002), Scarpa and Thiene (2005)). Furthermore, a distinct class (anyhow, close enough) is that of Generalized Nested Logit (GNL) models, Wen and Koppelman (2001). This class of models, which generalizes the Nested Logit (NL) model, impose an a-priori tree structure with nests and nodes. The relationship between the NL and the Multinomial Logit model (MNL) is very strong because an NL model can be viewed as the product of a series of MNL models, each MNL for each node. The main issue of the GNL model could be its flexibility due to the nesting structure; undoubtedly, this can also be viewed as a limit because an a-priori tree-structure must be imposed.

Finally, before entering into details related to Conjoint Analysis and Choice Experiments (Sects. 8.2.1 and 8.2.2), we briefly outline some features about Contingent Ranking. In this situation, the ordinal response variable conditions the respondent's interview (repeated and ordered choices) and, therefore, the statistical models to be apply. The repeated and ordered choices, called also panel, create a correlation between choices which can not be adequately treated through the Rank Ordered Logit (ROL) also when including the Alternative Specific Constant (ROL-ASC). An improvement may be obtained through the Kernel Logit (KL) model, which allows to take care of heteroschedasticity and correlations; in general, in this case, an Alternative Specific Constant (ASC) is introduced in order to discriminate, during the model estimation, for the status-quo (Herriges and Phaneuf (2002)). A recent study (Van Dijk et al. (2007)) introduces the concept of latent segments (Latent Class) jointly with a Rank Ordered Logit model (ROL-LC), in order to treat the heterogeneity of respondents due to their difficulty at ranking.

8.2.1 Choice Experiment: theory and advances

As shown hereinabove (Table 8.1), the CE theory considers the experiments and statistical models as main theoretical elements; nevertheless, further issues should be evaluated in order to completely discuss this methodology, such as the estimation methods and simulation algorithms to solve the model's expression, Bhat (2001). In this brief section, we mainly focus on the model step, by evaluating the solutions suggested in the literature in order to solve the effective problems when this method is applied.

The role of the experimental design is not irrelevant when we consider its properties; broadly speaking, the search of a D-optimal design implies the maximization of the determinant of information matrix and, therefore, this directly influences the variances of parameter estimates and, obviously, the volume of the ellipsoid, confidence region for the parameters, which is strictly connected to the precision of the design. This implies a larger efficiency in the estimates. In the specific literature about CE, the consideration of a D-optimal design, from the fractional factorials to more complex designs (Zwerina et al. (1996), Yu et al. (2009)) built through specific algorithms of trial-point selections, has been replaced in recent years by using Bayesian optimal designs (Kessels et al. (2004)). However, the experimental planning through D-optimal designs can not be considered as a limited tool because it guarantees optimal properties jointly with a notable manageability in comparison with the implementation of Bayesian designs.

By considering the experimental planning for a choice or conjoint experiment, some features are general common rules for a valid experimental planning, independently of the application field. Therefore, the attributes must be accurately defined in their number and in the number of levels. Surely, an experimental design formed by attributes with the same number of levels is more easy to treat; at the same time, a great attention must be paid to the distance among levels. Undoubtedly, the inclusion of a large number of attributes with distant levels increases the complexity of the design and the decision of the user/consumer becomes more difficult and implies a response error; Swait and Adamowicz (2001) face this problem from the point of view of the choice capability and its difficulty through a heteroschedastic Multinomial Logit Model.

In Scott (2002) a problem of dominant preferences is focused in the health care system, by considering the consumer's decision task and its complexity when evaluating the defined levels and the presence of a lexicographic preference- i.e. when the consumer always prefers the same alternative, independently of the other alternative settings. In addition, a relationship between a general alternative and the status-quo or current situation, is created according to these general criteria. If alternatives are very distant, a problem of a dominant alternative could be found; on the contrary, when alternatives have close level values, the presence of the status-quo alternative could be much more appealing and the respondent tends to prefer the current situation without changing. Nevertheless, the inclusion of status-quo alternative cannot be disregarded in Choice Modelling (CM) for the interpretation and estimation of economical concepts, first of all the Willingness To Pay (WTP) for a relative change in each single attribute. Furthermore, the complexity of choice and the planning issues outlined previously must be evaluated together with the number of choice-sets given to a single respondent and with the kind of response variable adopted, ranking or binary-choice variable. Undoubtedly, in the CR field, the complexity is increased by the ranking task; on the other hand, for example, in CE environmental situation, a choice-set is usually comprised of two alternatives and the status-quo alternative. In the literature, several studies attempted to improve these issues by starting from the planning phase or by considering improvements in the model step; in DeShazo and Fermo (2002) the sources of variability are studied in order to identify the impact of complexity on the consistency of choice, by introducing measures of complexity and studying the effect of complexity as in the variance-components field.

8 Choices and conjoint analysis

In fact, the authors analyze the problem by defining an heteroschedastic logit model according to the five complexity measures defined; thus, the dependent variable is the variability due to the characteristics of the choice-set. In a previous study, on the same subject, (Dellaert et al., 1999), the consistency is evaluated by considering the specific attribute of cost. Recent developments about the complexity of choice and related problems, such as discontinuity, where discontinuity could be defined as a "break point" in the likelihood function due to extreme situations of the consumer/user's behaviour, are studied in Gilbride and Allenby (2004), where discontinuity points are evaluated in the estimation step by introducing the concept of consideration-set and screening-rules for consumers. Thus, threshold values are defined by discriminating according to specific rules of consumer's utility. In Campbell et al. (2008) the impact of discontinuous preferences, from the point of view of respondents, is evaluated on the WTP estimates; the respondents with discontinuous preferences are identified during the decision process through a multinomial error component logit model which includes the constant term, namely the ASC, in order to consider the status-quo situation. These authors deal with the correlation between the utility of changing alternatives and the status-quo aspect together with the heterogeneity due to the different type of the respondent's preference. A very interesting remark is the consideration of different scale parameters according to the number of respondents' discontinuities; this allows to treat differently the sets of respondents owing to their preferences.

Furthermore, recent developments about the WTP estimates are in Garrod et al. (2002), Strazzera et al. (2003), Scarpa et al. (2007), Sonnier et al. (2007); in Scarpa et al. (2007) this theoretical problem is faced by defining a parallel Willingness To Pay (WTP) space where parameter estimates are evaluated by considering a more specific economic definition of the WTP, in order to improve its interpretation; in Sonnier et al. (2007) a Bayesian approach for the WTP estimates is introduced. Willingness To Pay estimates and zero values, according to the typology of response motivations, are studied in Strazzera et al. (2003).

In order to deal with the above features, the general class of Random Utility Models (RUM) is defined. In general, every alternative is indicated by j ($j = 1, ..., J$), while i denotes the consumer/user ($i = 1, ..., I$); thus, the following expression is characterized by a stochastic utility index U_{ij}, which may be expressed, for each unit i, as:

$$U_{ij} = V_{ij} + \varepsilon_{ij} \qquad (8.1)$$

where V_{ij} is the deterministic part of utility, while ε_{ij} is the random component, independent and Gumbel distributed. The class of RUM, which aims to achieve the utility maximization, enlarges the characteristics of Logit and Nested Logit (NL) models where the Independence of Irrelevant Alternatives (IIA) is hypothesized. The relaxation of this assumption is undoubtedly a very substantial improvement because the IIA means that the choosing probability in one choice-set is independent of the presence of other attribute values or any other alternative; on the other hand, we may say that IIA derives from the hypothesis of independence and homoschedasticity of the error terms. In addition, this can also be interpreted by considering the cross-elasticity term. In fact, IIA implies an equal proportional substitution between alternatives.

Furthermore, these models cannot take account of a different behaviour of the consumer; i.e. each respondent, with different baseline characteristics, is treated in a similar way (the same estimate values of attributes) according only to their judgement, exclusively.

In the literature, a first contribution to improving these issues is in Train (1998), where a Random Parameter Logit (RPL) model is introduced. At present, this model is more precisely called Mixed Multinomial Logit (MMNL). In fact, this RUM model allows to evaluate the respondents' heterogeneity or, better, the consumer/user's variability is estimated by considering the attributes as random variables and not fixed variables, i.e. as random variables across respondents; in addition, just because more choice-sets are supplied to the respondents, the repeated choices (during time) imply a correlation which is confounded with the consumer/user's variability (unobserved utility).

In this case, an appropriate example is in Train (1998) where, in a fishing case, the unobserved utility of the consumer is identified in the difference for each fisherman, when he must choose the fishing site. Further, according to repeated choices, this unobserved utility is confounded with the correlation due to several sites and trips; so, a correlation over trips and over sites for each fisherman's decision must be taken into account.

A general formulation for a single decision, according to McFadden and Train (2000), for a MMNL is:

$$Pr_C(i \mid \mathbf{x}; \boldsymbol{\lambda}) = \int_{\Re^I} L_C(i; \mathbf{x}, \alpha) G(d\alpha; \boldsymbol{\lambda}) \quad (8.2)$$

$$L_C(i; \mathbf{x}, \alpha) = \frac{\exp(x_i \alpha)}{\sum_{j \in C} \exp(x_j \alpha)}$$

where $C = (1, ..., j, ...J)$ is the general choice-set; $\mathbf{x}$ is the vector of attributes ($\mathbf{x} = x_1, ..., x_j, ...x_J$), and x_i is the observed value of the decision i; α is the vector ($Ix1$) of random parameters which expresses the respondents' heterogeneity. The term $L_C(\cdot)$ is the general expression for a Multinomial Logit (MNL), where the $G(\cdot)$ is the mixing component. It is very important to note that the random parameter α varies in the mixing term, where the differential over the integration is performed over α, because is $G(d\alpha; \boldsymbol{\lambda})$, where $\boldsymbol{\lambda}$ is the vector of parameters related to the mixing distribution.

By considering expression (8.2), we may further assume that an individual i belongs to the s group or segment, $(s = 1, ..., S)$, i.e. we assume a finite number of groups identified through the consumer baseline characteristics; from a theoretical point of view it is like assuming that the mixing term $G(\cdot)$ is defined on a finite support. Therefore, the probability for the unit i of belonging to the set s is included in $[0, 1]$ and $\sum_s Pr_{is} = 1$.

8 Choices and conjoint analysis

The deterministic term of the utility function may be expressed through a function of attributes and the characteristics of the s group; thus, the utility function defined by (8.1) may be now formulated as:

$$U_{ij|s} = V_{ij|s} + \varepsilon_{ij|s} \tag{8.3}$$

$$V_{ij|s} = \alpha_s x_{ij} \tag{8.4}$$

Note that formula (8.3) expresses the deterministic term conditional to the belonging to the group s and the specific choice is weighted through the utility characteristics of the set s.

Therefore, for each segment s, the probability to choose the alternative j^* for the unit i belonging to s is:

$$Pr_{i|s}(j^*) = \frac{\exp(\mu_s \alpha_s \mathbf{x}_{j^*})}{\sum_{j \in C} \exp(\mu_s \alpha_s \mathbf{x}_j)} \tag{8.5}$$

where α_s is the specific utility parameter for the segment s and μ_s is the specific scale factor, usually re-scaled to one, and here generically assumed. Note that if $\mathbf{x}_{j^*}$ is an alternative-specific value, as defined in Sect. 8.2.1, then α includes the alternative specific variable, in this case evaluated as a random effect.

Furthermore, we have the following relation:

$$Pr_i(j^*) = \sum_s Pr_{is} Pr_{i|s}(j^*) \tag{8.6}$$

Formula (8.6) expresses the global likelihood for a generic individual i who prefers j^* as the sum of products of two terms: the probability of the unit i of belonging to the group s is multiplied by the probability of the unit i belonging to s to choose the alternative j^*.

Formula (8.6) may be explicitly written as:

$$Pr_i(j^*) = \sum_s \frac{\exp(\beta \gamma_s \mathbf{z}_i)}{\sum_s \exp(\beta \gamma_s \mathbf{z}_i)} \frac{\exp(\mu_s \alpha_s \mathbf{x}_{j^*})}{\sum_{j \in C} \exp(\mu_s \alpha_s \mathbf{x}_j)} \tag{8.7}$$

where $\mathbf{z}_i$ is the vector of baseline individual characteristics, γ_s is the vector of parameters for the group s, while β is the scale factor, usually re-scaled to one as in (8.2) when considering L_C.

The last formula (8.7) could be interpreted as the model expression for a Latent Class Model (LCM) and it is also called as finite-mixture model (Boxall and Adamowicz (2002)), in comparison with the MMNL, formula (8.2), where the mixing term is assumed as distributed according to a continuous distribution (normal or log-normal, for example); by referring to formulas (8.2) and (8.7), in this case, the probability to choose the j^* alternative of the choice-set C is multiplied by the probability to choose given that i belongs to the group s.

It is not so irrelevant to remark that the scale factors μ_s should be posed equal to one in order to avoid imposing parameter values. This formula (8.7) expresses a flexible range of situations: if $S = 1$ there are not differences between baseline characteristics of the consumers; otherwise, if $S = I$ a group is defined for each unit, assuming the extreme situation of a total differentiation between individuals.

8.2.2 Conjoint Analysis: theory and advances

Conjoint Analysis (CA), (Netzer et al., 2008), can be defined, in our opinion, as the historical MAV method, where the term conjoint means the measurement of relative attribute values jointly. The first studies were made in 1970s where the basic fundamental theory of CA was posed (Johnson, 1974; Green and Rao, 1971; Green and Srinivasan, 1978). In Johnson (1974) trade-offs among alternatives were evaluated by considering a pair of attributes at-a-time and the respondent (consumer) must rank his/her preference as to these two attributes. The empirical example reported, (Johnson, 1974), is about the car-market, and the author assumes the independence of attributes and his analysis does not include the interaction first order terms. A not irrelevant point is a first introduction of the individual's characteristics, through a suggested weighting.

In Green and Srinivasan (1978), a further improvement was introduced, by considering a full profile (and not paired) evaluation where the global consumer/user's utility is then decomposed in order to estimate each single attribute's utility. Rating and ranking are the response variables preferred and the suggested statistical analysis usually applies the linear regression model. In the paired comparisons case, logit and probit are applied. Then, further studies have developed these issues, by pointing out model definition and estimation, (Green, et al., 1981; Green, 1984), where a hybrid utility estimation model for CA is suggested. Here, a self-explicated model, based on a procedure of measuring preference functions, is used with a conjoint model, with the inclusion of interaction (I order) terms. This model, which also takes account of differences (through clusters) of respondents, by evaluating their similarities in the self-explicated model, may be considered a basic model of CA. An enlargement of this model has different parameters for each attribute within each cluster. However, the correlation due to the evaluation of the same attributes in the two model steps is not completely assessed. A further note relates to the burden of respondent in this context; the respondent, in CA and more precisely for participating to a hybrid CA model, is asked to perform a heavy task, because he/she must participate two times to a judgement procedure.

Undoubtedly, the first attempts of respondents' segmentation are found in CA method. Currim (1981) and Moore (1980) applied strategies in order to satisfy the need of an intermediate level of consumer's aggregation through cluster analysis, individual a-priori information and preferences.

In fact, the multivariate statistical analysis played a relevant role in CA during 1980s, and the first half of 1990s (Punj and Stewart, 1983; Hagerty, 1985), above all the cluster analysis.

After dealing with this historical picture of fundamental CA elements, and, in general, with the evolution of MAV methods, we point out the different configurations of CA and the following developments in order to gain the respondent's coherency and reliability.

Starting from Green and Srinivasan (1990), CA had many differentiations according to a decompositional or compositional or mixed approach. All of these methods are related to an easy-to-treat multi-attribute situation; the self-explicated technique, just cited, requires the respondent to have a two-step evaluation; the Adaptive Conjoint Analysis (ACA) also applies the mixed approach and a paired comparison is performed through a computer-assisted interview. It is important to note that the attempts are directed towards a mitigation of the respondent's task, especially when the number of attributes and/or levels is high (Netzer and Srinivasan, 2007). In this respect, some studies are in common with Choice Modelling (CM) methods, such as De Bruyn et al. (2008).

A dynamic evaluation of CA was implemented in Bradlow et al. (2004); a consumer's learning phase is suggested through partial conjoint profiles in order to avoid the missing levels problem, which may exist when the experimental planning is conducted by a fractional factorial design, which is a reduced design of the corresponding full factorial design. Surely, optimal designs and specific algorithms are further solutions in order to overcome this problem.

In Bradlow et al. (2004), as in Lenk et al. (1996), the respondent's heterogeneity is taken into account through the application of a hierarchical Bayes model. Furthermore, in Lenk et al. (1996), the reduction of the number of profiles supplied to each respondent is studied in order to improve the estimation accuracy.

Therefore, three issues are variously combined in order to solve the CA problems: the reduction of the consumer's task; the complexity of data collection (the experimental planning step); the respondents' heterogeneity. Even though some features are strictly connected with CM methods, as was said previously, the respondent's stimulus, also introduced in Green and Srinivasan (1990), was also studied in recent years. Conjoint Analysis often appears in mixed techniques, such as in Barone et al. (2007) and in Schütte and Eklund (2005), where a Kansei Engineering (KE) is applied; KE is a multidisciplinary approach where the consumer is stimulated through real perceptions of existing products in order to give a weight to the technical and performance characteristics.

The link between consumer's preferences and the engineering field is largely used in recent years. The Response Surface Methodology (RSM) is also applied in the Customer Satisfaction (CS) field, see, for example Danaher (1997), but the strict relation between CA and the quality measures is in Kazemzadeh et al. (2008), Du et al. (2006) and Jiao and Tseng (2004). In Jiao and Tseng (2004), an Adaptive CA is applied jointly with the RSM. Two issues must be outlined: the concept of mass customization, i.e. the product design performed to attract the consumer's attention (through cost and customization value), and the consideration of a cost variable. A remark is the evaluation of RSM in a discrete context. An approach similar to Jiao and Tseng (2004) is in Du et al. (2006), where a more in-depth analysis is performed by considering the unit costs. Indexes about quality performance, costs and satisfaction are defined and applied in Kazemzadeh et al. (2008).

8.3 Our proposal: conjoint analysis and response surface methodology

The aim of the present study is the proposal of a modified Conjoint Analysis (CA) in order to establish an optimal solution for the product/service from the point of view of the user/consumer. The subsequent procedure is performed through the Response Surface Methodology (RSM) theory, by considering the quantitative judgement of each respondent for each profile with respect to the assessed score about the status-quo, and taking into account the individual information. The final result is achieved by carrying out an optimization procedure on the estimated models, and defining an objective function in order to reach the optimal solution for the revised (or new) service/product. Furthermore, it is relevant to point out the modifying structured data, through a new questionnaire, in order to collect information about the baseline variables of the respondent, the quantitative data about the current situation (status-quo) of the product/service, and the proper CA analysis by means of the planning of an experimental design. Therefore, two remarks must be made: the former is the consideration of the status-quo as the current situation for a revised product, otherwise the status-quo may be interpreted as the center of design or, alternatively, the full profile which identifies the medium situation; the latter is that the search of the best profile for the respondent is performed on a surface delimited by the range of attributes and centered on the status-quo.

8.3.1 The outlined theory

In this and in the following sections we briefly explain the RSM theory and the general optimization measures applied, according to a robust design approach (for details see Khuri and Cornell, 1987). Note that we focus our attention on the statistical models and optimization in the RSM; the fundamental elements of the experimental design (Box et al., 1978) are, however, indirectly introduced through the experimental planning related to CA.

In this case, the concept of a robust design approach is used for optimizing the service/product as more insensitive as possible with respect to the respondents heterogeneity or in order to adjust the service/product by considering those respondents characteristics which are relevant for the product/service studied. In general, we may define the set of experimental variables, which influence the measurement process: $\mathbf{x} = [\mathbf{x}_1, .., \mathbf{x}_k, .., \mathbf{x}_K]$ and the set of noise variables: $\mathbf{z} = [\mathbf{z}_1, .., \mathbf{z}_s, .., \mathbf{z}_S]$. In this context, the set $\mathbf{x}$ are the judgements, expressed through votes in a metric scale [0,100], on the attributes involved in the experimental planning; while the set $\mathbf{z}$ is related to the baseline individual variables, which are relevant for the service or product studied and that may change according to the specific situation. The response variable $\mathbf{Y}$ is defined as a quantitative variable of the process; in this case, the judgements expressed, on each full profile of the plan, by the respondents in the same metric scale. Note that, in general, if J are the profiles and I the respondents, the observations are $I \times J$. The general RSM model can be written as:

$$Y_{ij}(\mathbf{x},\mathbf{z}) = \beta_0 + \mathbf{x}'\beta + \mathbf{x}'\mathbf{B}\mathbf{x} + \mathbf{z}'\delta + \mathbf{z}'\Delta\mathbf{z} + \mathbf{x}'\Lambda\mathbf{z} + \mathbf{e}_{ij} \quad i=1,...,I; j=1,...,J \quad (8.8)$$

where $\mathbf{x}$ and $\mathbf{z}$ are the vectors of judgements attributes as described above; β, $\mathbf{B}$, δ, Δ, and Λ are vectors and matrices of the model parameters, $\mathbf{e}_{ij}$ is the random error which is assumed Normally distributed with zero mean and variance equal to σ. Λ is a $[K \times S]$ matrix which plays an important role since it contains the parameters of the interaction effects between the $\mathbf{x}$ and $\mathbf{z}$ sets.

In general, a noise variable may be defined as a categorical or quantitative variable which is also controllable and measurable. In the technological context, a noise effect which has these characteristics is introduced in the experimental design to reduce the pure experimental error and to set the variables controlling the process variability in order to find the experimental run which is the most insensitive to noise, through the first order interaction effect. In this situation, the set $\mathbf{z}$ is comprised of measurable categorical or quantitative variables which measure the baseline respondents characteristics. Therefore, the best profile is reached through the estimation of (8.8) conditional to the heterogeneity of respondents, taking into account judgements and individual data through the interaction terms. Furthermore, the response variable is comprised of the individual scores for each hypothetical profile and this information is used to gain an optimal solution on the surface around the status-quo (the attribute judgements $\mathbf{x}$) and conditionally to $\mathbf{z}$. In addition, it is not irrelevant to observe that the individual characteristics are an external source of variability with respect to an ideal design of service or product. In order to perform this procedure, it is necessary to effect a combined interview, with three steps. (1) gathering information about baseline variables; (2) quantitative judgements about each attribute in the status-quo when the service/product is revised, or, when the service/product is new, the judgements about each attribute in the medium profile: $\mathbf{x}_0 = (0,..,0)$; (3) the quantitative judgement on each full profile for each respondent. Note that the set $\mathbf{x}$ is the same by considering either the attributes involved in the experimental design (profiles) and the attributes in the status-quo.

Therefore, the prospective evaluation of the new or revised product/service is obtained by computing the optimal hypothetical solution through the status-quo. Note that, as explained hereinafter (Sect. 8.3.2), the estimated surface is subsequently optimized in order to gain the best preference on the basis of the attributes and judgments involved. Nevertheless, if the service/product studied is new, the status-quo is the centered scenario (always hypothetical); if the service/product studied is under revision, then the status-quo represents the real and current scenario in comparison with the other hypothetical scenarios supplied in the CA step, as usually happens in a Choice Experiments context (Sect. 8.2.1).

A further issue about the baseline variables must be outlined. In general, there are some aspects we wish to examine and which may have an influence on the expressed judgement of the respondent. We refer to those aspects related to social and demographic characteristics such as gender, age, educational level, income, job status. As described in Sect. 8.2, the heterogeneity of respondents plays a central role in MAV methods and this is confirmed by recent developments in the literature. In fact, there are sensible reasons to believe that such features affect the final results. In this proposal, we suggest and compare two analyses which differently include the individual information. In the first analysis, baseline variables are included in the model (8.8) as explained before; if a baseline variable is categorical, as the gender, this must be considered both in the estimation of the model and in the optimization step, carrying out an optimal surface for each level of the categorical variable.

This proposal is compared with the consideration of building a-priori strata according to the baseline level variables. In this case, the response surface model (8.8) does not include the set of variables $\mathbf{z}$ which are used to build the strata. The comparison is not trivial, just because in the first case we may estimate the interaction effects which may add useful information to obtain the full optimal solution; in the second case, where the problem of a categorical baseline variable is not relevant, the stratification allows to carry out the optimization process within every a-priori stratum.

8.3.2 The searching of the best profile through optimization

As was said in the previous section, our aim is the optimization of the the model (8.8) according to the status-quo situation. The expressed rate for each conjoint profile is considered as the response or dependent variable (formulated on a continuous scale); for example, a vote expressed according to the metric scale [0,100] may lead to a valid evaluation of the response as a continuous variable. Therefore, in general, the optimal target score may be defined as the maximum value of the metric scale; in the above example, this is equal to 100.

Two optimization measures are defined for the optimization process, with only one dependent variable; both measures allows to consider the optimization within a specific delimited surface defined by the range of attribute scores. The first measure is formulated by considering the quadratic deviation of the estimated surface model $\hat{Y}$ from the maximum score τ. Therefore, the formula to be minimized is the following:

$$F_1 = (\hat{Y}(\mathbf{x},\mathbf{z}) - \tau)^2 \qquad (8.9)$$

The second optimization measure is defined by considering the approaching of $E(\hat{Y})$ to the maximum score; thus, we carry out the minimization of the model variance of $\hat{Y}$ jointly with the approaching to the ideal maximum score.

The formula is:

$$minF_2 = V(\hat{Y})(\mathbf{x},\mathbf{z}) = E(\hat{Y} - E(\hat{Y}))^2 \qquad (8.10)$$

according to the following decomposition of the Mean Square Error (MSE):

$$MSE = E(\hat{Y} - E(\hat{Y}))^2 + (\tau - E(\hat{Y}))^2 \qquad (8.11)$$
$$B(\tau) = (\tau - E(\hat{Y}))^2 \qquad (8.12)$$

where (8.12) explains the adjustment of the expected score value to the target score. A further issue is about the computation of $E(\hat{Y})$ which is calculated by considering the expected value of the estimated surface model.

The optimization procedure, carried out through the Statistical Analysis System (SAS) and the procedure NLP, is preferably computed using non-coded data, just because we are not in a technological context (for further details, see Berni and Gonnelli, 2006). Note that, as regards optimization, the final result expresses the optimal score for each attribute involved in the model according to the respondents' preferences. Furthermore, the final optimal score for an attribute may be explained as the importance/utility of that variable in order to reach the best profile when considering the judgement of the respondent about the current situation (status-quo).
A further consideration may concern the inclusion of a categorical baseline information in the optimization process by including the proportion of units belonging to each level of the baseline variable, or belonging to level combinations for several baseline variables (strata), as in Robinson et al. (2006) where this case is studied in a technological field; nevertheless, the optimization must be always performed taking account of different surfaces according to these different strata.

8.4 Case study

The main aim is the evaluation of an interdisciplinary degree course of the University of Florence. As regards the data collection, a "questionnaire" is planned and submitted to a sample of students of the II-nd and III-rd year. The questionnaire is articulated on three parts according to the three different sets of information: (i) baseline variables; (ii) judgements about status-quo; (iii) the specific planned experimental design for the basic CA.

Every judgement is expressed on the metric scale $[0, 100]$. The first set of variables is related to the social and demographical data for each student: gender, age, exam average, enrolment status, job status. In the second part, the current situation is analyzed according to the specific five attributes: contents of the basic subjects (cb); practice/laboratory (pl); intermediate exam (ie); exam modalities (me); professional subjects for the future job (prof), see Table 8.2.

The third part contains the conjoint study planning through a fractional factorial design 2_V^{5-1}. Note that the profiles are 16 and the students are 46; therefore, the total number of observations is 736. Furthermore, in the following application, we consider as noise categorical variable the job status of the student (job), identified also by case (i) working, and case (ii) non-working.

Table 8.2 Attributes and levels

Attributes	−1	1
cb	basic subjects with lower theoretical deepening	basic subjects with higher theoretical deepening
pl	practice and laboratory as compulsory part of typical courses	practice and laboratory only as two distinct courses
ie	one intermediate exam	no intermediate exam
me	oral test with practice	written and oral test
prof	a general degree course in order to continue studies	a more specific degree course, in order to seek a job

8.4.1 Optimization results

The general response surface model (8.8) is applied by considering judgements of the full profiles and judgements on the attributes in the current situation. Parameter estimates with standard error and p-values are displayed in Table 8.3. Note that all the variables are significant, except "prof", which has a non significant p-value. However, this main effect must be inserted given that it is relevant when considering the interaction effects of "prof" with the other variables, and, above all, with the "ie" variable. In addition, a highly significant p-value results for the interaction effect of "prof" with the noise variable "job". The same observation can be made considering "ie" and "cb". The optimization procedure is performed applying the two measures (8.9) and (8.10) defined above. The optimization results are described also by considering diagnostic results such as: the objective function value (of), the infinity norm of the gradient ($\|x\|_\infty$), the determinant of the Hessian matrix ($|\mathbf{H}|$). We have also checked the max-step, i.e. a specified limit for the step length of the line search algorithm, during the first r iterations. Two surfaces are optimized, according to the two levels of the job variable: working and non-working. The results are shown in Tables 8.4 and 8.5, related to the results about the measures (8.9) and (8.10), respectively. We must point out that, in this case, even though convergency is always reached and diagnostic results are quite satisfactory, the starting diagnostic results are not perfect. The reason of this problem may be leaded to the kind of data, so different with respect to technological data, where the experimental trials are usually conducted with high accuracy.

In this respect, we must remark that a non controllable variability due to the respondent is implicitly inserted in our data. In fact, in this context, the optimization measure (8.10) is more precise with respect to measure (8.9) just because the computation of $E(\hat{Y})$ takes care of non orthogonal data and of moments values. This is also confirmed when selecting the best fitted models; in this context, by including or not a model term may be very relevant for the following optimization procedure. By considering the optimization measure (8.10), the best solution considers "cb" and "ie" (Table 8.5) as relevant attributes for the non-working students. The scores are very high for case (ii): 84.98 and 99.90, respectively. The attribute "ie" is included in the final solution also for the working students. The scores for the opti-

8 Choices and conjoint analysis

Table 8.3 Model estimates; job status as noise variable

Parameter	Estimate	Stand. Error	t-value	p-value
Intercept	−281.20	47.863	−5.87	0.0001
cb	2.33	0.751	3.10	0.0020
pl	−1.54	0.705	−2.18	0.0293
me	3.05	0.843	3.62	0.0003
ie	4.04	0.667	6.05	0.0001
prof	0.36	1.069	0.34	0.7375
job	−44.53	24.862	−1.79	0.0737
cb*pl	−0.01	0.004	−2.23	0.0262
cb*ie	0.02	0.005	4.43	0.0001
cb*me	−0.05	0.011	−4.17	0.0001
cb*prof	0.01	0.008	1.30	0.1958
cb*job	−2.69	0.359	−7.49	0.0001
pl*ie	−0.03	0.004	−5.76	0.0001
pl*me	0.067	0.012	5.41	0.0001
pl*prof	−0.01	0.006	−2.99	0.0029
me^2	0.02	0.007	3.76	0.0002
me*ie	−0.08	0.011	−7.02	0.0001
me*prof	−0.02	0.011	−1.91	0.0565
me*job	0.24	0.154	1.54	0.1232
ie^2	0.01	0.004	1.31	0.1909
ie*prof	0.02	0.005	3.76	0.0002
ie*job	−1.04	0.218	−4.79	0.0001
prof*job	3.52	0.430	8.18	0.0001

Table 8.4 Optimization through measure (8.9). Case (i) working; case (ii) non-working

Results	measure (8.9); case (i)	measure (8.9); case (ii)
Best score:cb	cb = 26.76	cb = 7.01
Best score:pl	pl = 0.00	pl = 2.00
Best score:me	me = 0.00	me = 4.00
Best score:ie	ie = 34.50	ie = 46.09
Best score:prof	prof = 0.00	prof = 59.47
o.f.	5.0e-27	2.0e-28
$\|x\|_\infty$	8.6e-13	1.3e-13
$\|H\|$	< 10e-8	< 10e-8

Table 8.5 Optimization through measure (8.10). Case (i) working; case (ii) non-working

Results	measure(8.10); case (i)	measure (8.10); case (ii)
Best score:cb	cb = 42.04	cb = 84.98
Best score:pl	pl = 0.00	pl = 0.00
Best score:me	me = 0.00	me = 0.06
Best score:ie	ie = 56.85	ie = 99.90
Best score:prof	prof = 0.00	prof = 22.10
o.f.	3.2e-27	3.2e-25
$\|x\|_\infty$	7.5e-13	1.0e-11
$\|H\|$	<10e-8	< 10e-8

mization measure (8.9) show very low values for all variables involved, except "ie" and "prof" in case (*ii*), Table 8.4. This may be viewed as a higher interest of non-working students versus professional learning. As regards case (*i*), (Tables 8.4 and 8.5), "cb" and "ie" are the only relevant attributes; however, in table 8.4, scores are low for both variables, 26.76 and 34.50 respectively; while, by considering measure (8.10), "ie" and "cb" achieve higher scores (42.04 and 56.85, respectively). These solutions allow us to hypothesize a larger consideration of the professional elements by the non-working student in comparison with the one who works. The optimal solution obtained through measure (8.10) highlights the importance of "cb", "ie", "prof", by confirming the results obtained applying the (8.9) and the previous considerations about the relevance of computing $E(\hat{Y})$.

Furthermore, we compare these results with those obtained by using the baseline variables, in particular the job status of the student, for setting a-priori strata.

Two response surface models are estimated within each level of the job variable (estimates are not shown); "prof", "cb" and "pl" are significant attributes for the working students. The estimated surface model related to the non-working students allows us to confirm a large interest towards "cb" and "prof". Furthermore, "prof" is a common relevant attribute within each stratum; "pl" is relevant when considering the working students, while "cb" is more relevant for the students without a job, which express a great interest towards the basic courses in conjunction with more professional tools.

By considering the optimization results, (Tables 8.6 and 8.7), the diagnostic measures are always good, even though the results obtained through measure (8.10) have a high objective function value; however, the values of $|\mathbf{H}|$ are very good. Optimization measures (8.9) and (8.10) highlight "pl" and "prof" as the attributes with the highest scores for the working students. Within non-working students, "prof" and "cb" result as relevant attributes, confirming the propensity of the non-working student towards studying.

Table 8.6 Optimization through measure (8.9); a-priori strata; case (i) working; case (ii) non-working

Results	measure (8.9); case (*i*)	measure (8.9); case (*ii*)
Best score:cb	cb = 0.13	cb = 65.77
Best score:pl	pl = 73.57	pl = 0.00
Best score:me	me = 0.00	me = 0.00
Best score:ie	ie = 0.00	ie = 0.00
Best score:prof	prof = 84.21	prof = 50.99
o.f.	8.1e-28	1.5e4
$\|x\|_\infty$	1.3e-13	2.1e-14
$\|H\|$	<10e-8	8.4e-1

Table 8.7 Optimization through measure (8.10); a-priori strata; case (i) working; case (ii) non-working

Results	measure (8.10); case (*i*)	measure (8.10); case (*ii*)
Best score:cb	cb = 0.00	cb = 65.77
Best score:pl	pl = 56.49	pl = 0.00
Best score:me	me = 0.00	me = 0.00
Best score:ie	ie = 0.00	ie = 0.00
Best score:prof	prof = 100.00	prof = 51.00
o.f.	2.7e4	3.4e2
$\|x\|_\infty$	2.1e-9	3.6e-15
$\|H\|$	<10e-8	1.9e-2

8.5 Concluding remarks

By concluding, the main feature of this empirical example is the application of RSM jointly with CA in order to establish the best profile according to the judgements, expressed in metric scale, on the full profiles and on the status-quo. With this approach it is possible to take into account both a new service/product and a revised one. In addition, baseline variables of respondents, evaluated as noise variables, are introduced in the optimization procedure, by also considering their categorical nature. Note that in this case (Sect. 8.4.1) an only one surface is estimated and two optimal solutions are evaluated in the optimization step. The empirical results confirm the relevance of our proposal, also when comparing these results with the optimization within a-priori strata and the working situation.

Chapter 9
Robust diagnostics in university performance studies

Matilde Bini, Bruno Bertaccini and Silvia Bacci

> *In almost every true series of observations, some are found, which differ so much from the others as to indicate some abnormal source of error not contemplated in the theoretical discussions, and the introduction of which into the investigations can only serve to perplex and mislead the inquirer.*
>
> (Peirce, 1852)

9.1 Introduction

The presence of anomalous observations (*outliers*) in a set of data is one of the greatest problems in methodological statistics, one that scientists were already aware of many years ago, as can be seen in the comments made by the American astronomer Peirce[1] over 150 years ago.

An anomalous value can be defined generally as an *observation that appears to be non-compatible with the probabilistic model that has generated the rest of the data* (Barnett and Lewis, 1993). However, this statement bears a certain degree of subjectivity the moment the judgement of compatibility has to be expressed. This concept can be explained if we consider certain divorce cases that have been filed on the grounds of abnormalities discovered in the duration of pregnancies (Barnett, 1976). In 1949 a certain Mr. Hadlum appealed against the rejection of his previous

---

Matilde Bini
Department of Statistics "G. Parenti", University of Florence, Viale Morgagni 59, 50134, Florence, Italy, e-mail: bini@ds.unifi.it

Bruno Bertaccini
Department of Statistics "G. Parenti", University of Florence, Viale Morgagni 59, 50134, Florence, Italy, e-mail: brunob@ds.unifi.it

Silvia Bacci
Department of Statistics "G. Parenti", University of Florence, Viale Morgagni 59, 50134, Florence, Italy, e-mail: s.bacci@ds.unifi.it

[1] Peirce was the first to test for the objective identification of anomalous observations.

case for divorce, which was based on his wife's suspected adultery because she had given birth to a child 349 days after he, the husband, had left to go to the war. Since the average gestation in humans is 280 days, 349 days seemed surprisingly long to Mr. Hadlum, causing him to judge the span of time an anomalous value, that is, an observation deriving from another population (in this case, originating after the moment declared by the wife). Mr. Hadlum lost the case; the Court of Appeal maintained that the wife's period of gestation, while highly improbable ("extreme" value), was not scientifically impossible, in contrast to Mr. Hadlum who considered the value "contaminant", that is, deriving from a different distribution and therefore clear evidence of the wife's adultery. It is very likely that Mr. Hadlum would not have been suspicious if, for example, his wife had given birth 290 days after his departure for the war: nevertheless, even this length of time for the gestation could have been considered a "contaminant" value in the afore-mentioned sense, since the wife could have become pregnant, for instance, 20 days after the husband's departure plus a normal gestation of 270 days.

The keywords used here are *outlier, extreme value and contaminant value*. Therefore, these concepts must be made clearer, and an attempt must be made to formalise their definitions; for this purpose, we must consider an ordinate sample of n univariate observations

$$x_{(1)}, x_{(2)}, \ldots, x_{(n)}$$

all deriving from a certain F distribution, with the exception of two from a G distribution (see Fig. 9.1).

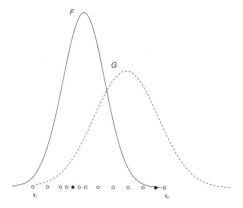

Fig. 9.1 Outliers, extremes values and contaminants.

The $x_{(1)}$ and $x_{(n)}$ observations are the extreme values of the sample; in the case being examined, however, only $x_{(n)}$ can be considered anomalous because of its position in relationship to the F model generator hypothesized. Hence, the extreme values are not necessarily *outliers*, whereas any individual *outlier* is always an extreme (or relatively extreme) value of the sample. The observations deriving from the G distribution are indicated with a black dot in the figure; though both can be defined contaminant values, only the second one, coinciding with $x_{(n-1)}$, can be considered anomalous regards F, differing from $x_{(n)}$ (even more anomalous than $x_{(n-1)}$), which nevertheless is not contaminant. Hence, the contaminant values can be or cannot be identified as *outliers*, and these can be more or less contaminant values (that is, deriving from distributions different to the one hypothesized). Unfortunately, there is no way we can know if an observation is contaminant (this is why Mr. Hadlum would probably not have been suspicious of a pregnancy lasting 290 days); all we can do is try to understand *if the outliers are possible manifestations of some form of contamination* (Barnett, 1988).

The importance of this problem led Box and Andersen (1955) to coin the term "robust" when referring to estimation methods that continue to have desirable properties, in spite of the fact that part of the data might result presumably contaminated to a certain degree.

In this respect, Tukey (1960) defines the mixture $(1-\varepsilon)F + \varepsilon G$ a contaminated distribution, where the F distribution is contaminated by the G distribution with ε probability (known as *contamination quota*). In his famous work, for evaluating the effects of a casual-type contamination on the efficiency properties of traditional estimation procedures, Tukey presumed the extraction of a sample of n observations from the contaminated distribution

$$(1-\varepsilon)N(\mu,\sigma^2) + \varepsilon N(\mu,9\sigma^2).$$

Figure 9.2 illustrates the effect of the contamination Tukey proposed for certain ε values between 0 and 0.5: the contamination weighs down the tails of the original distribution, and this weight becomes heavier as the ε value increases.

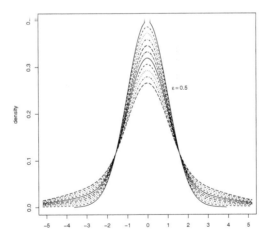

Fig. 9.2 Tukeys' mixture.

Tukey demonstrated how small, natural contaminations (between 1 and 10%) in the theoretic model could make the traditional asymptotic theory on optimality absolutely insignificant.[2]

There are various sources of contamination that can produce anomalous values; sometimes they are concomitant and they are certainly never known beforehand. The anomalous aspect might reflect the natural variability of the phenomenon being investigated, generated by erroneous measurement or, more often, they might derive from mistakes in the "implementation of the design" caused by distraction or due to ignorance of the person responsible for recording the information.[3] But detection of these sources appears as a problem of secondary importance compared to the adoption of tools that allow efficacious identification of the *outliers*.

After a description of the characteristics of robust methods in Sect. 9.2, an introduction of the Forward Search algorithm and its implementation in the Generalized Linear Models applied to the university effectiveness evaluation are reported in Sect. 9.3. Section 9.4 presents the Forward Search for the fixed effects ANOVA models and its application to the evaluation of the Italian University reform. Finally, Sect. 9.5 is devoted to some concluding remarks.

9.2 Robust methods vs diagnostic analysis

The aim of any diagnostic analysis is to define a more or less general outline of the collective phenomenon under investigation, in order to identify the peculiar characteristics, which, at a further stage, will require advanced statistical techniques and tools. This process of comprehension often originates in the implementation of a *probabilistic model*,[4] capable of synthesizing the state of knowledge of the phenomenon in question. Therefore, the model is a description of the process that generates the data and its implementation becomes the main criterion for recognition of the anomalous observations; hence, a non-typical value is considered such if it is able to produce a "surprise" effect regarding the particular probabilistic model presumed to have generated the data. For instance, in a sample of 7 observations made up of:

$$0.47, 6.18, 0.09, -0.60, -1.09, -1.19, 1.86$$

the second value is surprising if connected with a theory that presumes a probabilistic model of the N family $(0,1)$ to be the generator; however, this supposition

---

[2] In his famous work, Tukey proposed a comparison of relative efficiency between the $d_n = \frac{1}{n}\sum |x_i - \bar{x}|$ and $s_n = \left[\frac{1}{n}\sum(x_i - \bar{x})^2\right]^{1/2}$ estimators of the σ variability parameter, revealing how only two out 1,000 observations are able to annihilate the efficiency of the s_n estimator; in particular, Tukey demonstrated how d_n results to be preferable for all the ε values within the interval [0.002, 0.52].

[3] The mistakes in measurement and in "implementation of the design" (which lead to the inclusion of non-representative units) are defined by Anscombe (1960) as false observations.

[4] The description of any type of reality is often a very complicated operation, because of the interrelationships that are nearly always present.

would appear completely misleading since the data have been really generated by a Cauchy distribution with a parameter of 1 on the scale.[5]

From a typically parametric point of view, the initial theories indicate a probabilistic model in this manner, hoping it to be only a fair approximation of reality, without ever being able to presume that it is absolutely correct. Hence, all statistic procedures should have the following, desirable properties (Huber, 1981):

1. they should demonstrate a reasonable (almost optimal) level of efficiency regards the model presumed;
2. they must be robust, meaning that slight deviations from the theoretic model should bring about similarly slight penalization in performance (for instance, the asymptotic variance of an estimator ought to be near its nominal value as calculated in relationship to the theoretic model);
3. any appreciable deviation from the theoretic model should not cause a "catastrophe".

At this point, one might wonder if the robust procedures are really necessary or if, on the other hand, it would be sufficient to resort to the traditional procedures after adopting some technique that can discard the anomalous observations.

Unfortunately, this is not the case. First of all, the techniques that discard outliers are not free of errors; in Tukey's example, the removal of *outliers* from the dataset generated by mixed distribution would continue to produce a sample of observations that are not normal – because of wrong exclusions and wrong preservation of data – leading to a *framework* that would be just as inappropriate as the initial one, which advises against applying the traditional theory to normality.

Moreover, the difficulties encountered in the detection of anomalous values increase as the number of variables composing the structure of the available data rise. In fact, a univariate analysis of the context, while being an important part of the statistical procedure, is often of limited interest since many modern investigational techniques (confirmed by appropriate graphical analyses) are able to distinguish atypical situations.

Much more interesting is the multivariate analysis, in which the spatial composition of the observations makes the placing of anomalies less intuitive and, consequently, the formal methods for detecting them much more complex.

Lastly, the best procedure for removing *outliers* cannot match the performance shown by the best robust procedure Barnett and Lewis (1993). Indeed, this latter is definitely superior because it is a gradual (and not immediate) transaction between the total acceptance and the total deletion of an anomalous observation manifesting a contaminant distribution.

In the case of linear regression models, the presence of anomalous values can be easily depicted by simply *plotting* the data or the residuals; however, even in the multiple regression model, when the number of explicative variables increases, their detection by means of graphical tools may not be so immediate, especially in the presence of groups of *outliers* that mask each other.

---

[5] Observations beyond the body of the data can be caused by casual extractions from the Pareto and Cauchy distributions.

To overcome this problem, some estimation methods called *robust* or *resistant* have been proposed; these terms are used in the literature to illustrate their capacity to produce estimates that are not easily influenced by contaminant data. These methods all identify as *outliers* those units that show the highest residuals. Among the various *robust* approaches proposed, special mention must be made of the Least Median of Squares – LMS – (Rousseeuw, 1984) because it is intuitive and easy to use. However, the robust estimators (*LMS, MAD, trimmed mean*, etc.) have the disadvantage of under weighing or neglecting some of the observations; furthermore, they can fail completely if the observations do not derive from one population alone, but from various distinct populations.

Another approach to the problem is through the so-called *diagnostic analyses*, which foresee statistical calculation capable of detecting the anomalous values and the most influent among them. These can be examined and then either deleted or corrected, in order to allow the model to re-adapt by means of the traditional techniques. Worthy of note among the diagnostic techniques is that known as the *single deletion diagnostic*, which, at every step of the analysis, foresees the elimination of one observation at a time from the n available, and calculation of the new estimates and new parameters on the remaining observations. With two *outliers*, pairs of observations can be deleted and the process can be extended to several units at the same time. However, the traditional diagnostic methods suffer from serious inconveniences; the masking effect that takes place in the presence of groups of *outliers* makes the individual influence of each single one very limited and therefore unidentifiable; this aspect requires the diagnostic process to be extended to several observations simultaneously. Nevertheless, one realises immediately that the combinatory explosion of the number of observations to be taken into account can create considerable problems from a computational point of view and in interpretation of the results. An alternative to these limitations can be to repeat the single deletion processes; but in this case, the set of observations used in the adaption process decreases as the analysis proceeds. Atkinson and Riani (2000) demonstrated that this procedure, which is commonly known as *backward deletion*, might sometimes fail.

Note that the *robust* approaches, in spite of the fact that they focus on the same diagnostic target, proceed in a completely opposite manner, adapting the model first of all by using techniques that take into account the characteristics of the dataset, then examining the units that diverge most from the predicted values. However, the two approaches often lead to the same results. Some authors, while agreeing on the need to resort to robust criteria for the analysis, do not approve of the deletion of cases that have been really observed (though many robust methods do not consider the *outliers* at all); on the contrary, others, in spite of agreeing on the necessity to remove the anomalies, maintain that to resort to a robust method rather than another type is arbitrary (even though the preventive deletion of the observations and the subsequent adaption by means of ordinary least squares is itself a robust method).

In effect, this debate does not at all solve the problem of the *outliers*. What is undoubtedly important is to judge each single technique on the basis of the number of *outliers* it manages to identify – or tolerate – before they can influence the inferential process somehow. This property is formalised by introducing the *breakdown point* (Donoho and Huber, 1983; Rousseeuw and Leroy, 1987) which is defined the smallest fraction of contamination that can make a certain estimator assume values far away from the estimates that would have been obtained if the contamination was absent.[6]

9.2.1 The Forward Search algorithm

The *Forward Search* (Atkinson and Riani, 2000; Atkinson et al., 2004) is a procedure that is capable of combining the efficiency of traditional inferential methods with the capacity to identify anomalous observations within a sample of data and then assess the effects achieved. Its main feature is that it proceeds in a manner exactly opposite the *backward* one, that is typical of the traditional diagnostic methods that assess the anomaly or the influence of an observation on the statistic model only after this has been adapted to the entire sample of data. On the contrary, the *Forward Search*, given the n observations of the sample, starts by searching among the data available for a minimal dataset presumed – on the basis of the model – to be free from *outliers*. This starting subset is detected by means of different approaches according to the analysis context: in the case of linear regression models, the adaption of a high number of small subsets is evaluated, employing robust statistical methods to define which of these procedures produces the best adaption; in the case of multivariate statistical analysis, *boxplot* bivariate matrix and *spline* functions adapted to the actual placement of the observations in space are used.

The evolution of the procedure is therefore ensured by evaluation of the adaption to increasingly larger observations obtained by the sequential inclusion of the remaining observations in relationship to their proximity to the theoretic model. The process obviously stops when all the units observed participate in the inferential process. The arrangement of the data performed at every step excludes the problems of masking encountered by the traditional diagnostic methods, the targets of which are reached by monitoring the various statistics (i.e. the goodness of fit test, significance of parameters) during the evolution of the algorithm.

The result of this procedure is the arrangement of the observations with respect to the degree of their proximity to the presumed model[7]; in the case of linear regression, this arrangement is achieved by starting from a robust adaption and reaching one of ordinary least squares.[8] Monitoring the various statistics usually employed in the traditional inferential approaches permits gathering a set of data capable not only of detecting the *outliers* but also – and this is even more important – of understanding the influence that each of them has on the inference of the model.

The next sections propose the robust diagnostic analysis made using the forward search algorithm as a useful tool to detect anomalous situations in university performance evaluation.

---

[6] For example, for ordinary least squares even one observation alone might be sufficient for this to take place. In this case, the OLS estimator would have a breakdown point of 0%.

[7] In regression models, "proximity" is expressed by the residuals; in multivariate analysis, by a measurement of distance (Mahalanobis, Manhattan, etc.).

[8] If the model agrees with the data, the robust adaption and the least squared one will produce similar results, both in estimation of the parameters and in the errors. However, the estimates and the residuals of the adapted model change considerable during the search process.

9.3 The Forward Search for Generalized Linear Models

The Forward Search is an approach for detecting the presence of outliers and assessing their influence on the estimates of the model parameters. The method was first applied to regression analysis, but it could as well be applied to almost any model (Atkinson et al., 2004). The procedure starts out by fitting the model to a subset of the observations, say m observations, which is chosen in some robust way. The observations of the entire set are then ordered by their closeness to the estimated model. The model is then refitted using the subset of the $(m+1)$ observations which are closest to the previously estimated model. The observations are ordered again, the model is refitted to a larger subset and the process is continued until all the data have entered. At every step the subset size is increased by one unit (usually one case is added to the previous subset, but sometimes two or more are added as one or more leave the subset), bringing about an ordering of all the observations. At every step, the fitting of the model will also produce estimates of the parameters of the model under study as well as other relevant statistics. Changes of these statistics, as the Forward Search is carried out, are analyzed (graphically or otherwise) for the purpose of assessing the influence of each observation on the estimation of the model and - under the hypothesis that the outliers are the last ones to enter - of identifying a cut-off point that divides the outliers from the "good" data. More formally, it is based on the following steps:

The start is a robust fit to very few observations and then a successive fit is done with larger subsets. The initial subset is identified using the *least median of squares method* (Rousseeuw, 1984) that guarantees that no outliers are included in the initial subset.

Formally, (see details in Atkinson and Riani, 2000, p. 31): let $Z = (X, y)$ a data matrix of dimension $n \times (p+1)$. If n is moderate and $p << n$ the choice of the initial subset can be performed by exhaustive enumeration of all $\binom{n}{p}$ distinct ptuple $S^{(p)}_{i_1,\ldots,i_p} \equiv \{z_{i_1},\ldots,z_{i_p}\}$, where $z^T_{i_j}$ is the *ij*th row of Z, for $j = 1,\ldots,p$ and $1 \leq i_j \neq i_{j*} \leq n$.

Specifically, let $\iota^T = [i_1,\ldots,i_p]$ and let $e_{i,S^{(p)}_\iota}$ be the least squares residual for the unit i given the model has been fitted with the observations in $S^{(p)}_\iota$. The initial subset is $S^{(p)}_*$ which satisfies

$$e^2_{[\text{med}],S^{(p)}_*} = \min_\iota \left[e^2_{[\text{med}],S^{(p)}_\iota} \right] \qquad (9.1)$$

where $e^2_{[k],S^{(p)}_\iota}$ is the *k*th ordered squared residual among $e^2_{i,S^{(p)}_\iota}$, with $i=1,\ldots,n$ and med=integer part of $(n+p+1)/2$. If $\binom{n}{p}$ is too large, the choice is made using 3,000 ptuples sampled from Z matrix.

The subset size is increased by one and the model refitted to the observations with the smallest residuals for the increased subset size.

The initial subset $S_*^{(m)}$ of dimension $m \geq p$ is increased by one and the new subset $S_*^{(m+1)}$ consists of $m+1$ units with the smallest ordered residuals $e^2_{[k], S_*^{(m)}}$. The model is refitted to the new subset and the procedure continues increasing subset sizes until all the data are fitted, i.e. when $S_*^{(m)} = S^{(n)}$.

The result is an ordering of the observations by closeness to the assumed model.

9.3.1 Robust GLMs for the university effectiveness evaluation. The case of the first year college drop out rate

One of the most important indicators of efficacy, as well as of efficiency, which the Ministry takes into account in judging the teaching activity of a university, is the *drop out rate*, which continues to be extremely high in all the Italian universities even after the reform. Recent findings confirm that the drop out rate is still well over 30 percent and this calls for new research to discover the reasons and possible solutions of a problem that has strong social and political implications.

Past research and commonsense tell us that the most important factors affecting the probability of dropping out from college are the characteristics of students at time of enrollment, but also possible changes of their characteristics during their study, as well as the characteristics of teaching activity. These factors are important for every university but their impact is probably different for different institutions. This justifies a research on drop out rate conducted on data from the University of Florence (Italy) (Bini et al., 2003; Bini and Bertaccini, 2007).

The present study has been performed using two sets of data which have been linked. Administrative data, collected by each Italian university at time of enrollment and survey data, collected in June 2003 through Computer Assisted Telephone Interview (CATI) interviews of the students that enrolled at the University of Florence in the year 2001–2002.

The analysis was conducted using regression models which attempted to explain the probability of dropping out by using a set of individual, institutional and contextual variables.

Since the observed response variable is a dichotomous one, the estimation of such probabilities were first made through generalized linear models with classical estimation procedures (maximum likelihood estimation). This was considered as an exploratory phase of the project which would identify a set of significant variables as well as an improved general understanding of the problem. The data were then fitted using a robust approach proposed by Atkinson and Riani (2000), whose results will be reported and commented in another publication.

The procedure we adopted for the estimation of the model classifies all the observations with a hierarchical order in terms of adherence to the model. The characteristics of the extreme groups will be analyzed with descriptive methods and, hopefully, will give us information which could be useful for planning policies that will reduce the university drop out rate.

The use of the robust approach allows identifying singles or groups of students with particular characteristics, for example it may be possible to find groups of individuals who withdrawn the same course programs, otherwise single or groups of freshmen who leaved different course programs. In the first case the explanations should be given to the characteristics of these course programs: it is probable that the learning is too difficult because of the capability and behaviour of instructors, or the organization of classes is poor, or some other reasons due to the teaching activity. The second case, that is groups of withdrawers of different course programs, could depend to specific characteristics of these freshmen (because they are workers, or they got a low score at high school), or even the information about them is biased due to the interviews not correctly carried out or the questionnaire not so clear.

9.3.1.1 Dataset descritption

The analysis on dropout, regarding to all the freshmen enrolled in the past 2001–2002 a.y., used a data set containing some information from the administrative data and some other ones collected from a survey conducted by the Department of Statistics of the University of Florence in June 2003. A number of 2,908 freshmen who left the initial attended course program, which represent the 30% of the total freshmen of that year (10,053 cases). From questionnaire the first information we have, is about the different kinds of withdrawal (here called *profiles of drop out*) like moving from one to another course program, or degree program, or even to another university, withdrawal declared with a written communication or not declared.

Then, for all the withdrawers we know:

1. the reasons of changes due to the university activity, such as problems concerning:

   - the organization of the structures (i.e. if classrooms, laboratories, libraries are adequate to number of students and to technology requested for teaching, etc...);
   - the organization of the course (i.e. amount of class hours with respect to the length of semester, schedules, number of exams during the semesters, etc..);
   - teaching quality of instructors (i.e. clarity, instructor enthusiasm, usefulness of exercises, organization of examinations, the availability of teachers after class, workhome, materials of study, etc...;

2. personal reasons such as: change of residency; health problems; family problems; the occupational status at the enrolment.
   Moreover:
3. in case of moving to another course, degree program or university, which degree and course program they enrolled;
4. in case they did not enroll, or there is no news about their enrolment, whether they intend to enroll again, and eventually in which course/degree program and university.

9 Robust diagnostics in university performance studies

The response variable of interest in these regression models is the students drop out, identified with the binary variable Y as follows: Y = 1 if dropped out includes all the profiles of withdrawal except to one concerning students who moved to another course or degree program; Y = 0 otherwise.

A preliminary descriptive analysis revealed that among all the variables included in the updated administrative data set, only the covariates shown in Fig. 9.3 yielded a strong association with the response variable.

However, just two covariates are strictly linked to the response variable as indicator of the efficacy and efficiency of teaching activity (i.e. *Course Program* and *Degree Program* selected at the enrolment); whereas the other ones strictly pertain to characteristics of individuals, i.e. gendre (Sex), Age at the enrolment (AgeEnroll), Residence (County), Kind of High school (Hschool), Level score of undergraduate entrants (HSScore), Occupational status when enrolled (Occup).

Covariate	Description	Levels
Sex	Gender	1 = 'Male'; 2 = 'Female'
AgeEnroll	Age at the enrolment	1 = '< 20'; 2 = '20'; 3 = '21 - 25'; 4 = '>25'
County	Residence	1 = 'Florence Hinterland'; 2 = 'others municipality counties without Florence and Prato'; 3 = 'Other provinces of Tuscany; 4 = 'Other regions of northern and central Italy'; 5 = 'Other regions of southern Italy and islands'
Degree	Degree program selected at the enrolment	1 = 'Agriculture'; 2 = 'Architecture'; 3 = 'Economics'; 6 = ' Pharmacy'; 7 = 'Law'; 8 = 'Engineering'; 9 = 'Letters & Philosophy'; 10 = 'Medicine'; 11 = 'Educational Science'; 12 = 'Political Science'; 13 = 'Psychology'; 14 = 'Maths & Physics'; 15 = 'Interdisciplinary'
Course Program	Course program selected at the enrollment	104 levels
Hschool	Kind of High school	1 = 'Classical'; 2 = 'Scientific'; 3 = 'Technical'; 4 = 'Others'
HSScore	Level score of undergraduate entrants	1 = '60 - 56'; 2 = '55 - 51'; 3 = '50 - 46'; 4 = '45 - 41'; 5 = '40 - 36'
Occup	Occupational status when enrolled	1 = 'Employed'; 2 = 'Unemployed'

Fig. 9.3 Covariates with a stronger association with the response variable.

9.3.1.2 Fitting models

The fit of a logit model for binary data on this reduced data set, allowed to select the significant covariates (AgeEnroll, County, Degree, Hschool, HSScore).

The aim of this study is to detect groups of students, having particular characteristics.

To this purpose the analysis is accomplished by grouping individuals on the basis of the levels of the significant covariates.

The result of this grouping yielded some very low size clusters among 826 groups obtained, so that, as it is known, some test statistics are not reliable anymore. Then, a second fit of a logit model followed using a data set formed by a number of clusters equal to 454, each one at least having more than five units.

The related results led to reject the *kind of high school* covariate (see results in Fig. 9.4).

```
Model: binomial, link: logit
Response: y
Terms added sequentially (first to last)

            Df    Dev.  Res.Df  Res.Dev    P(>|Chi|)
NULL                       453  1126.89
County       4   61.71    449  1065.18    1.267e-12
Degree      12   81.14    437   984.04    2.506e-12
AgeEnroll    3  383.22    434   600.82    9.932e-83
HSScore      3  105.09    431   495.73    1.249e-22
```

Fig. 9.4 Fitting logit model for binomial data.

9.3.1.3 Main results

The algorithm of forward search applied to GLMs, has been carried out with a macro implemented using the R package.

Looking at some forward plots, it is possible to observe the influential importance of some groups.

The first plot to be consider is the goodness of link test along the forward search.

This plot allows to explore different possible link functions (logit, probit, cloglog and arcsin) and then to choice the best one. Looking at the t statistic values of each link put together in a plot (not reported here), it can be noted that trajectories are different and the order of introduction of the observations is different for each link. Moreover, each statistic has an increasing or decreasing trend and it goes outside of bounds (at the 5% level) at different steps of the forward search. Although this means a bad fit in all these cases, from a comparison among trends, it follows that the arcsin link is the most satisfactory one.

Figure 9.5, which reports the goodness of link test of the arcsin function, shows decreasing trend of the statistic as we move towards the end of the forward until it goes out of the significant bounds (5% level) after the step $m = 356$.

9 Robust diagnostics in university performance studies 151

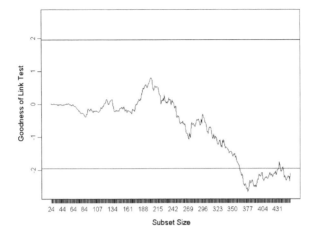

Fig. 9.5 Forward Search: goodness of link tests.

This is due to the presence of groups that differ more than other ones from the bulk of the data; more specifically, they are the last 98 groups entering the subset (as highlighted by the red circle). The presence of observations different from the bulk of data, as well as their affecting the fit of the model, is also highlighted from the monitoring of the deviance of the model.

In Fig. 9.6 it can be noted that the last 98 clusters entering the subset (after the step $m = 356$) cause an exponential increasing of the values of the residual deviance and an exponential decreasing of the values of the Pseudo-R2 statistic; this means that these observations have a significant influence.

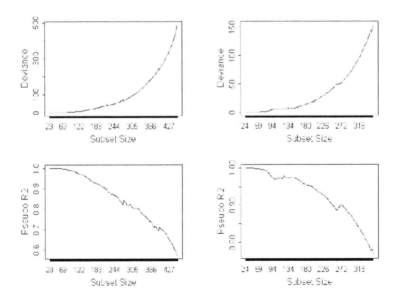

Fig. 9.6 Forward Search: deviance of the Model.

The importance of the effect of the influential groups can be well depicted, once again, by plotting the values of the goodness of link test (5% level). Figure 9.7 reports the three plots of the t statistic during the forward search respectively with the entire data set, after deleting the last 5 and then the last 98 clusters entered the subset. Even though the statistic is inside the bounds after the deletion of the last 98 groups, the trend still shows the bad fit of the model due to all the observations.

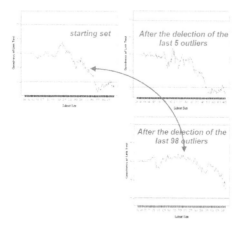

Fig. 9.7 Forward Search: goodness of link tests.

Once the groups of outliers have been detected, the next step should be the investigation of the characteristics of the units inside the groups by the implementation of descriptive analyses, that should allows us to depict various situations useful for the intervention policies aiming to improve the teaching quality and consequently to reduce the drop out rate.

As an example of this kind of analysis, let consider, here, two particular situations arose last steps of the search: the first and the third last groups entering the subset, respectively labeled 270 and 62.

Cluster composition & characteristics

Course	Obs	Variable	Mean
Nursery (9.2% of the cluster)	11	ageEnroll	32.54
		HighSchoolScore (36-60)	37.90
		Y (drop rate)	81.8%

Note: Freshmen enrolled from Tuscany but out of the counties of Florence

Course characteristics

Course	Obs	Variable	Mean
Nursery	197	ageEnroll	21.73
		HighSchoolScore (36-60)	42.13
		Y (drop rate)	27.4%

Note: 31% of Freshmen enrolled from Tuscany but out of the counties of Florence

Cluster composition & characteristics

Course	Obs	Variable	Mean
Various courses of Mathematical Science	47	ageEnroll	18.9
		HighSchoolScore (36-60)	58.9
		Y (drop rate)	0.0%

Note: Freshmen enrolled from the hinterland of Florence

Degree characteristics

	Obs	Variable	Mean
Mathematical Science	109	ageEnroll	21.09
		HighSchoolScore (36-60)	45.27
		Y (drop rate)	19.3%

Note (only freshmen from the hinterland of Florence - 32% of the total enrolled in this degree)

Fig. 9.8 Descriptive analyses of last cluster, number 270 (*left panel*) and of the third last cluster number 62 (*right panel*) entered the subset.

Results reported in Fig. 9.8 show their main characteristics. As concerns the cluster 270, it consists of 11 graduates of nursery, and about all the students (81.8%) dropped out.

It can be noted that with respect to the characteristics of students of the entire nursery course which shows a low drop out rate (27.4%) even including the outlier cluster, they are in average older and less "clever" as highlighted by the lower average of the high school score; moreover, as concerns the residence characteristic they all come from Tuscany region but out of the county of Florence, with respect to the 31% of the entire course program.

This particular situation tells us that the drop out of this course maybe depends more on the characteristics of students than to those of the teaching activity. Anyway, the supplementary information we have from survey, allows us to verify whether this conclusion is correct or some other reasons due to the courses efficiency have to be included.

The second example, indeed, depicts an opposite situation where 47 students attending different courses programs of Mathematical Science do not drop out (0%) and they have in average better performance than the average of all the graduates of the same degree program. About the residence characteristic, all students of the cluster live in Florence and hinterland, while only the 32% of the total enrolled in this degree (also here the number of 109 includes the 47 students of the cluster 62) have the same characteristics.

Here, since the drop out rate of the degree is quite low, we are led to not investigate on the characteristics of these courses but rather on the characteristics of students.

In this particular case, it should have even been interesting to collect, by specific interviews, the evaluations on teaching activities of the courses these outliers attend.

9.4 The Forward Search for ANOVA models

In this section the implementation of the Forward Search method in the ANOVA framework is presented, in order to identify the observations that differ from the bulk of the data and to analyse their effect on the estimation of parameters and on inferences on the model. The methodology is adapted to the peculiarity of the ANOVA models taking into account the differences between the fixed effects and the random effects ANOVA models. In particular, a procedure is explained to obtain a Robust Forward F Test for the former case and a Robust Forward Likelihood-Ratio Test (LRT) for the latter case.

We remind briefly the main characteristics of the one-way ANOVA model. Let y_{ij} be the observed outcome variable of individual i ($i = 1, 2, \ldots, n_j$) within group, or factor level, j ($j = 1, 2, \ldots, J$) where J is the total number of groups and $N = \sum_{j=1}^{J} n_j$ is the total number of individuals. The simplest linear model in this framework is expressed by:

$$y_{ij} = \mu + u_j + e_{ij} = \mu + x_{ij} \qquad (9.2)$$

where μ is the grand mean outcome in the population, u_j is the group effect associated with unit j and e_{ij} is the residual error at the lower level of the analysis. This model can be interpreted as a fixed or random effects model, depending on the assumptions about the nature of u_j. When u_j are interpreted as the effects attributable to a finite set of levels of a factor that occur in the data, we have a fixed effect model. On the contrary, when u_j are the effects attributable to a infinite set of levels of a factor of which only a random sample are deemed to occur in the data, we have a random effects model.

Classical assumption on the fixed effects ANOVA model is:

$$e_{ij} \sim N(0, \sigma^2) \ \forall i, j.$$

For the random effects ANOVA model other assumptions are added:

$$u_j \sim N(0, \tau^2) \ \forall j$$
$$cov(e_{ij}, e_{i'j'}) = 0 \ \forall i \neq i' \text{ and } j \neq j'$$
$$cov(u_j, u_{j'}) = 0 \ \forall j \neq j'$$
$$cov(e_{ij}, u_j) = 0 \ \forall i, j$$

and, as a consequence:

$$var(y_{ij}) = var(u_j) + var(e_{ij}) = \tau^2 + \sigma^2$$
$$cov(y_{ij}, y_{i'j}) = \tau^2 \ \forall i \neq i',$$

where τ^2 expresses the variance among groups and σ^2 expresses the variance within groups.

In the following sections the specific steps to implement the Forward Search are briefly reminded and then an application of the proposed approach to real data, using a set of information referring to the performance of the Italian university system is illustrated. For more details about the theoretical aspects of Forward Search for ANOVA models see Bertaccini and Varriale (2007) and Bertaccini and Varriale (2008).

9.4.1 The Forward Search for the fixed effects ANOVA

9.4.1.1 What is the problem in presence of outliers?

In the fixed effects ANOVA model, we are usually interested in the null hypothesis:

$$H_0: u_1 = u_2 = ... = u_J = 0,$$

that means that there is not any effect of the factor on the average level of Y.

The statistics used to verify the null hypothesis is defined as:

$$F = \frac{DB/(J-1)}{DW/(N-J)}, \tag{9.3}$$

where DB is the deviance between groups and DW is the deviance within groups. From the normality in the ANOVA assumptions, if the null hypothesis is true, it follows that the ratio statistics F is distributed as a Fisher's F distribution with $(J-1)$ and $(N-J)$ degrees of freedom.

Due to the presence of the sample means in both the DW and DB, the value of the F statistic is strongly affected by the presence of outliers. In fact, it is known that the sample mean is the best unbiased estimator of a population mean under normality assumption, but it shows a strong loss of efficiency in case of contamination or misspecification of the model. This means that in the presence of contaminated data, the "real" value of the first type error probability is systematically higher than the α nominal value (e.g. 0.01, 0.05, ...) and, therefore, the test F will often erroneously reject the null hypothesis.

9.4.1.2 The Forward Search steps

The methodology proposed takes into consideration the presence of groups in the data structure of the ANOVA model. At every step of the Forward Search parameters estimates, residuals, classical F value and other considerable statistics are computed. As usual in the Forward Search method, the procedure is carried on through the classical three steps, that are specified in according to the characteristics of the model:

- Step 1: choice of the initial subset

The specific proposal in the ANOVA framework is to start with the observations y_{ij} that satisfy $min|y_{ij} - med_j|$ in each group j ($j = 1,...,J$), where med_j is the group j sample median.

- Step 2: adding observation during the search

At each step, the Forward Search algorithm adds to the subset the observations closer to the previously fitted model. This can be accomplished following two different strategies: the first, called non-proportional, adds just one new unit at each step, while the other, proportional, enters the minimum number of observations necessary to respect the overall composition (the group proportions) of the sample.

- Step 3: monitoring the search

At each stage of the search, parameter estimates, residuals and other relevant statistics, such as classical F test values, are calculated in order to detect the outliers. The main difference between the non-proportional and the proportional approach is that, with the non-proportional strategy the observation belonging to the groups with the minimum variance will enter before the others: hence, the outliers will enter the model last. Instead, in the proportional strategy outliers are forced to enter together with good observations in order to maintain the proportionality of the dimension of the groups.

Finally, a Robust Forward F Test is defined to divide the group of outliers from the other observations, and to evaluate correctly the null hypothesis of fixed effects ANOVA model. The Robust Forward F Test can be defined as a collection $F_{FS} = F(k),...,F(n)$ of the classical F test in each step of the search; to obtain a Robust Forward F Test it is possible to individuate a cut-off point of the progress procedure dividing the group of observations that differ to the bulk of the data from the others. The search of the cut-off point can not be "automatic" but is completely based on graphical analysis and is strictly connected to the context of the observed phenomenon.

With the proposed method, the probability of accepting H_1 when H_0 is true is always lower than the same probability obtained with the classical ANOVA F Test.

9.4.2 The Forward Search for the random effects ANOVA

9.4.2.1 What is the problem in presence of outliers?

In a random effects model, such as the ANOVA ones in Eq. (9.2), the observations are aggregated in different levels, so that it is possible to discern first-level units and second-level units or groups. Therefore, we have two different kinds of outliers: first- and second-level outliers. For example, if we consider the hierarchical structure of university system, where students (or first-level units) are aggregated in degree programmes (or second-level units or groups), we could observe one or more students in one or more degree programmes that are anomalous with respect to the student population for some characteristics, or we could observe one or more degree programmes that are anomalous with respect to the degree programmes population. So, we need to focus on the evaluation of the effect of both first and second level outliers on the inferences on the model and, in particular, on their effect on the higher level variance which is statistically evaluated with the LRT.

In many applications of hierarchical analysis, one common research question is whether the variability of the random effects at the group level u_j is significatively equal to 0, namely

$$H_0 : \tau^2 = 0.$$

If the null hypothesis is accepted, then we can conclude that the hierarchical structure of data has no effect on the dependent variable Y. The most used procedure to test this hypothesis is the Likelihood-Ratio Test. In a random effects one-way ANOVA model the asymptotic distribution of the Likelihood-Ratio statistic is a mixture of Chi-squares distributions. However, due to the presence of outliers in the data, the value of the *LRT* statistic can erroneously suggest to reject the null hypothesis H_0 even when there is no second level residual variability.

Therefore, in presence of contaminated data, the classical LRT for the random effects ANOVA model has a similar behavior to the classical F test for the fixed effects ANOVA model. The "true" α value is systematically higher than the nominal ones.

9.4.2.2 The Forward Search steps

The three steps of the Forward Search for the random effects ANOVA model develop in a very similar way to the fixed effects ANOVA model.

- Step 1: choice of the initial subset

As in the previous case, the search starts with the observations y_{ij} that minimize $|y_{ij} - med_j|$ ($j = 1, ..., J$), where med_j is the group j sample median. Moreover, we impose that every group has to be represented by at least two observations; in this way, every group contributes to the estimation of the within random effects.

- Step 2: adding observation during the search

At each step of the search, all the observations are ordered inside each group according to their squared total residuals. The total residuals express the closeness of each unit to the grand mean estimate, making possible the detection of both first and second level outliers. For each group j we choose the first m_j ordered observations and add the one with the smallest squared residual among the remaining.

- Step 3: monitoring the search

At each stage of the search, parameter estimates, residuals and other relevant statistics, such as classical LRT values, are calculated in order to detect the outliers.

Among the most useful outputs there are the plots of the within ($\hat{\sigma}^2$) and between ($\hat{\tau}^2$) variance components and the values of the classical LRT estimated at each step of the Forward Search. See Varriale and Bertaccini (2009) for an analysis of the different trend of the two kinds of plots in presence of first-level outliers or in presence of second-level outliers.

Finally, a Robust Forward LRT is defined to evaluate correctly the null hypothesis of random effects ANOVA model. It can be defined in an analog way to the Robust Forward F test: it is a collection of the values of the classical LR Test statistic computed at each step of the search, and to obtain a Robust Forward LR Test we identify a cut-off point of the progress procedure that best divides the group of observations that differ to the bulk of the data from the others. With the proposed method, the probability of accepting H_1 when H_0 is true is always lower than the same probability obtained with the classical LRT.

9.4.3 The use of the robust ANOVA for the evaluation of the Italian university reform

In this section it is presented an application of the Forward Search for ANOVA models in order to evaluate the impact on the Italian university system of the reform on degree programs, that was enacted in the academic year 2001/02. One of the main aims of this reform was obtain a reduction of the withdrawal rate. The data come from annual surveys conducted by the Italian National University Evaluation Committee (NUEC) during the years 2001, 2002, 2004 and 2005 and refer to the activities of all the public universities during the academic years 1999/2000, 2000/2001, 2002/2003, 2003/2004; the study is limited to the Italian degree programs in Mathematical Science. The dataset is composed by four groups identified by the years in which the NUEC surveys were conducted: two years before the reform and other two after the reform, with 276, 283, 342, 351 observations, respectively. For our purposes, we use the first-year retention rate indicator (RR), defined as $RR = 1 - WR$, where WR is the withdrawal rate.

To find out the effect of the reform on the RT over different years a fixed effects ANOVA model is estimated, where RT is the dependent variable. First of all, a classical ANOVA F test is conducted: the F value is equal to 2.50 with a p-value equal to 0.058, that is larger than the nominal α value of 0.05. Therefore, on the base of the classical ANOVA F test, the null hypothesis is accepted, and we can conclude that the reform had no effect on the first-year retention rate.

Because of the presence of many outliers in the data set, it is interesting to conduct also a Robust Forward ANOVA F Test, in order to evaluate if the presence of outliers influence the results of the classical test. Among the outputs produced by the Forward Search procedure, the most interesting are shown. In Figs. 9.9, 9.10 and 9.11 are plotted the residual standard error, the estimated RRs, and the classical F values, respectively, at each step of the Forward Search.

The exponential increasing of the curve in Fig. 9.9 confirms the presence of outliers in the dataset, and we can see that they enter the model during the last steps of the Forward Search. From Fig. 9.10 it can be seen that the estimated RRs referring to the two years after the reform (2004 and 2005) are almost systematically higher than the two others. Only when the outliers enter the model the F values for years 2004 and 2005 converge to the F values referred to years 2001 and 2002. Finally, form the analysis of Fig. 9.11 is evident that the F statistics is always in the reject region: only when the outliers enter in the procedure, the F values fall in the acceptance region.

Therefore, on the basis of the results shown in Figs. 9.10 and 9.11 it can be concluded that: (i) by taking into account only data conformed with the ANOVA model hypotheses, the university system reform had a positive effect on the increasing of the first year retention rate; (ii) the presence of outliers induces to an opposite, and wrong, conclusion. This example highlights the superiority of the Forward Search approach in comparison with a classical approach, such as classical ANOVA F test, for the inference in presence of outliers.

9 Robust diagnostics in university performance studies 159

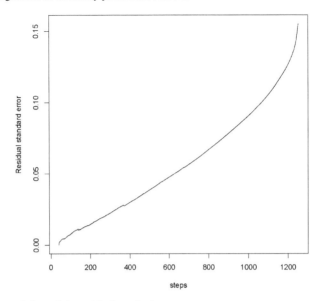

Fig. 9.9 Forward plots of the residual standard error.

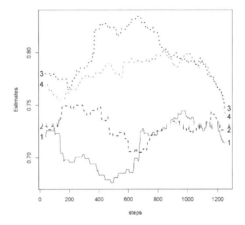

Fig. 9.10 Forward plots of the estimated coefficients for the first year retention rate of the Italian degree programs in Mathematical Science.

9.5 Concluding remarks

The peculiarities of *Forward Search* and its analytical possibilities, which have been highlighted in this work, make it an approach that is usually preferred to other robust methods; in fact:

1. *Forward Search* combines robustness and efficiency because, during the evolution of the analysis, the estimation procedures are based on well-known statistical algorithms (maximum likelihood, least squares,...) with proven efficiency

and quick computation abilities; in other words, no ad hoc high intensity computation algorithms are required for estimating parameters;
2. The approach is easily extended to different analytical contexts (regression, generalised linear models, multivariate method of analysis, etc.) and is therefore applicable to most of the situations of which multidimensional data are available;
3. The method can be generalised even to cases in which there is auto-correlation between the observations (historical sets, models for spatial data);
4. The approach features a higher degree of generality compared to other robust methods, since the *outliers* are neither deleted nor "underweighted"; *Forward Search* actually allows their entrance probably in the final stages of the procedure, thus offering the analyst the possibility of evaluating the effects on the inferential conclusions drawn from the adapted statistical model.
5. Finally, we know that the analyses and the representation of a certain degree of the performance of university are a useful support for planning some interventions and actions as concern the organization of the structures, but especially the teaching activities. To perform a deeper studies about the complex system of relationships and factors which affect problem like for example the drop out of University, or the impact of a new reform, it is necessary to use appropriate analytical models. The robust diagnostics regression analyses are able not only to supply an answer to these needs, but also allow identifying observations or groups of units (outliers) having specific characteristics. The inspections on these outliers could really help to implement university programmes on the teaching activities aimed at improving the quality of this service.

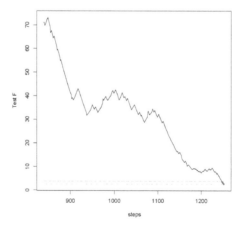

Fig. 9.11 Forward plots of the F statistics.

Chapter 10
A novel global performance score with an application to the evaluation of new detergents

Stefano Bonnini, Livio Corain, Antonio Cordellina, Anna Crestana, Remigio Musci and Luigi Salmaso

10.1 Introduction

In the research and development of new products often the aim is focused at evaluating the product performances in connection with more than one aspect (dimension) and/or under several conditions (strata). In this framework the main goal of statistical data analysis consists in the calculation of an index to obtain a global performance evaluation of the products under investigation which is a synthesis of the information given by whole performance data.

The goal of this chapter is twofold: at first we present a novel Global Performance Score (GPS) for the construction of a global performance index when we are facing a complex problem of product quality evaluation; then, we wish to investigate the main consequences for GPS when different standardization methods

---

Stefano Bonnini
Department of Mathematics, University of Ferrara, Via Machiavelli 35, 44100 Ferrara, Italy, e-mail: bnnsfn@unife.it

Livio Corain
Department of Management and Engineering, University of Padua, Stradella S.Nicola 3, 36100 Vicenza, Italy, e-mail: livio.corain@unipd.it

Luigi Salmaso
Department of Management and Engineering, University of Padua, Stradella S. Nicola 3,36100 Vicenza, Italy, e-mail: salmaso@gest.unipd.it

Antonio Cordellina
Research & Development, Reckitt-Benckiser Ltd, Piazza S. Nicolo' 12/3, 30034 Mira (VE), Italy, e-mail: Antonio.Cordellina@Reckittbenckiser.com

Anna Crestana
Research & Development, Reckitt-Benckiser Ltd, Piazza S. Nicolo' 12/3, 30034 Mira (VE), Italy, e-mail: Anna.Crestana@Reckittbenckiser.com

Remigio Musci
Research & Development, Reckitt-Benckiser Ltd, Piazza S. Nicolo' 12/3, 30034 Mira (VE), Italy, e-mail: Remigio.Musci@Reckittbenckiser.com

and aggregation techniques are used. The considered experimental design presents a multivariate response variable where the univariate components have different degrees of importance. In general, each dimension of the global performance should be evaluated under different conditions which can be represented by two or more strata, jointly considered. The methodological solution to cope with this problem is described and applied, considering different possible data transformation and an application problem related to the performance evaluation of new detergents.

Let us suppose that the global performance is represented by a variable η, that indicates a complex and underlying concept, often named *construct*, which is not directly measurable, hence it is broken into a set of measurable components, dimensions or items. In order to build up a global performance index, two main critical steps have to be taken into account: *standardization* and *aggregation*.

Standardization methods should take into account both the data properties and objectives of the analysis. Let $Y_1, Y_2, \ldots, Y_K$ be the informative variables representing the measurable components of η. Standardization of $Y_1, Y_2, \ldots, Y_K$ is a transformation that replaces each Y_k by a new variable $T_k(Y_k)$. The main goal of standardization is to allow for the comparability among variables. A review of the most commonly used transformations and an exploration of the main mathematical and statistical consequences of their application is proposed in Aiello and Attanasio (2004).

After the transformation of non homogeneous data, it is necessary to put together the variables $T_k(Y_k)$ through an *aggregating function* $g(\cdot)$. Hence the *aggregation* allows for obtaining a global final variable which gives a measure of the *construct* or *latent variable* η:

$$Y = g\left[T_1(Y_1), T_2(Y_2), \ldots, T_K(Y_K); \omega_1, \omega_2, \ldots, \omega_K\right], \qquad (10.1)$$

where $\omega_1, \ldots, \omega_K$ are the weights (degrees of importance) assigned to $Y_1, \ldots, Y_K$, respectively. Weights usually have an important impact on the aggregated values of the performance index. Although some weights could be negative, in general $\omega_k \geq 0$, $k = 1, 2, \ldots, K$ and $\sum_k \omega_k = 1$. From now on we will assume this condition unless a different assumption is explicitly done. The most frequent aggregation functions proposed in the literature (see Fayers and Hand, 2002, for an extensive review) are based on additive methods and require assumptions about indicators and weights which are often not desirable and difficult to meet and to test (Nardo et al., 2008). For this reason other aggregation methods have been proposed. Among these, we mention multiplicative methods, such as geometric aggregation, and multi-criteria analysis.

In this chapter we are facing the problem of determining a comparative global performance evaluation of C products, summing up partial performance measures coming from multivariate experimental data in presence of multistratification. The complexity of the experimental design is due to the following aspects: (i) the response variable is multivariate and the univariate component variables present different degrees of importance; (ii) one or more component variables represent primary performances, while other ones represent secondary performances and two partial aggregated evaluations are at least needed along with the global evaluation; (iii) some experiments (in general those related to primary performances) allow replications hence, for some responses, comparative evaluations can be based on multiple comparisons of one-way ANOVA but, for economic or practical reasons, other responses are characterized by unreplicated designs, hence, for these variables, inferential procedures are not possible; (iv) each dimension of the global performance should be evaluated under different conditions which can be represented by two or more strata, jointly considered.

The present chapter is organized as follows. Section 10.2 is dedicated to the description of the most common standardization and aggregation techniques which are devoted to properly define composite indexes. In Sect. 10.3 the general procedure to define and calculate the GPS is shown. In Sect. 10.4 the results of the application of GPS to a real industrial problem, using different normalization and aggregation methods, are presented. Section 10.5 illustrates a comparative simulation study where the combination of normalization and aggregation methods are compared. Finally, Sect. 10.6 is dedicated to the final remarks.

10.2 Composite indexes

A composite index has to measure a complex and underlying phenomenon η which is not measurable but can be broken into K measurable components, dimensions or items. Data transformation procedure for the calculation of a composite indicator consists in a sequence of steps aimed to achieve comparability among component variables $Y_1, Y_2, \ldots, Y_K$ and to make a synthesis of the available information. The former purpose is obtained through *standardization*. The latter can be achieved through the application of an *aggregation* technique.

10.2.1 Standardization: data transformations to obtain homogeneous variables

Let us suppose that Y_{ck} represents the value of k-th variable for c-th unit (product). A possible standardization approach to have comparability is to rank each variable across units. For example, in the case of decreasing *rank transformation* we have:

$$T_k(Y_{ck}) = R(Y_{ck}) = \sum_{u=1}^{C} I_{uk}(Y_{ck}) + 1, \qquad (10.2)$$

where

$$I_{uk}(x) = \begin{cases} 1 & \text{if } Y_{uk} > x \\ 0 & \text{otherwise} \end{cases}$$

This typical non linear transformation requires just simple calculations and is robust in presence of outliers. The main disadvantage is the loss of information related to the original metric. The evaluation of a unit based on a given variable consists in the position of the unit in the ranking based on that variable. Relative rank $R(Y_{ck})/C$ can be preferable to $R(Y_{ck})$ because it takes values in the interval $[0,1]$. To avoid computational problems in the aggregation phase (i.e. null denominator, null argument of logarithm, etc.), relative rank $[R(Y_{ck}) + c_1]/(C + c_2)$ can be calculated, where c_1 and c_2 are constants such that the relative rank takes values in the open interval $(0,1)$.

The traditional *standardization* method converts all original variables to variables with zero mean and standard deviation equal to one, applying the well-known transformation:

$$T_k(Y_{ck}) = \frac{Y_{ck} - \overline{Y}_k}{S_k}, \qquad (10.3)$$

where $\overline{Y}_k$ is the sample mean and S_k the sample standard deviation of Y_k. This transformation is not robust with respect to outliers. Sometimes a similar linear transformation, with median instead of mean as a location measure and median absolute deviation instead of standard deviation as a variability measure, is used.

The *re-scaling* technique produces standardized variables with identical range $[0,1]$:

$$T_k(Y_{ck}) = \frac{Y_{ck} - \min_u(Y_{uk})}{\max_u(Y_{uk}) - \min_u(Y_{uk})}. \qquad (10.4)$$

Since this method uses range instead of standard deviation as denominator, outliers have a great effect on standardization. Standardization can be applied just comparing the original data with the maximum value, according to the following formula:

$$T_k(Y_{ck}) = \frac{Y_{ck}}{\max_u(Y_{uk})}.$$

In this case the standardized variables assume values in $\left[\frac{\min_u(Y_{uk})}{\max_u(Y_{uk})}, 1\right]$. A similar transformation can be obtained just considering the minimum value, for instance:

$$T_k(Y_{ck}) = 1 - \frac{\min_u(Y_{uk})}{Y_{ck}},$$

and the range of standardized variable is $\left[0, 1 - \frac{\min_u(Y_{uk})}{\max_u(Y_{uk})}\right]$. It is worth noting that the latter transformation is nonlinear while standardization based on maximum value and (10.4), are linear.

When standardization is aimed at the *comparison with a reference unit* (or *with a target*), value one is given to the reference unit, i.e. $T_k(Y_k^*) = 1$, and transformed values are calculated through the ratio

$$T_k(Y_{ck}) = \frac{Y_{ck}}{Y_k^*}, \tag{10.5}$$

where Y_k^* indicates the value of Y_k corresponding to the reference unit or the target value for Y_k. With this method, typical of economic applications where all Y_{ck} are nonnegative (e.g. index numbers), transformed data take value in $[0, \infty)$. Alternatively, when we are interested in gaps, the relative variations $T_k(Y_{ck}) = (Y_{ck} - Y_k^*)/Y_k^*$, taking values in $(-\infty, +\infty)$, can be calculated.

A similar, but more robust, method of standardization distinguishes among values above, close to, or below a certain percentage threshold around the mean or a reference value:

$$T_k(Y_{ck}) = \begin{cases} -1 & \text{if } Y_{ck} - m_k < -\delta_k|m_k| \\ a_k Y_{ck} & \text{if } -\delta_k|m_k| \leq Y_{ck} - m_k \leq +\delta_k|m_k| \\ +1 & \text{if } Y_{ck} - m_k > +\delta_k|m_k| \end{cases}, \tag{10.6}$$

where m_ks are the means or the reference values, δ_ks are the percentage thresholds and a_ks are non negative constants. As a special case some a_ks could be null. The disadvantages of this nonlinear transformation are the arbitrariness of δ_k and the loss of the information about the original metric.

10.2.2 Aggregation: synthesis of information

The application of an aggregation technique consists in the choice of an appropriate function $g : \mathfrak{R}^K \to \mathfrak{R}$ to apply (10.1). The most used are additive techniques but they require assumptions which are often not desirable and sometimes difficult to meet and to verify (see Nardo et al., 2008). Hence some authors propose alternative aggregation methods such as multiplicative (geometric) aggregations or non-compensatory aggregations (e.g. multi-criteria methods).

Additive aggregation is based on the weighted sum of standardized variables:

$$Y = \sum_{k=1}^{K} \omega_k T_k(Y_k). \tag{10.7}$$

There should be no conflict or synergy among standardized variables. In case of conflict, standardization should be used also to change direction of the original variables decreasingly related to the latent variable Y.

Additive aggregation is a fully compensatory approach because low values in some variables can be completely compensated by sufficiently high values in other variables. Assuming all T_k are positive, *geometric aggregation* presents less compensability because it is based on a multiplicative approach:

$$Y = \prod_{k=1}^{K} T_k(Y_k)^{\omega_k}. \tag{10.8}$$

If a geometric aggregation is applied to calculate a composite performance indicator, a unit under evaluation should prefer to increase partial indicators (variables) with low score than those with high score to improve its position in the global ranking.

The *multi-criteria approach* is based on a non compensatory rationale. The main assumption is the comparability between units for each variable Y_k. The method consists in ordering the units after pair-wise comparisons across the whole set of variables. Multi-criteria analysis allows us for considering jointly qualitative and numeric variables and in general it does not necessarily require standardization to assure comparability among variables. For each variable Y_k a *preference function* is defined, such that for each couple of units (u,v), it indicates if u is worse, equivalent or better than v ($u,v = 1,\ldots,C$). The preference function can be written:

$$h_k(Y_{uk}, Y_{vk}) = \begin{cases} -1 & \text{if } u \text{ is worse than } v \text{ according to } Y_k \\ 0 & \text{if } u \text{ is equivalent to } v \text{ according to } Y_k \\ +1 & \text{if } u \text{ is better than } v \text{ according to } Y_k \end{cases} . \quad (10.9)$$

The above general definition of the preference function can be applied to a wide range of functions. The choice about which function should be used depends on the decision-making problem and from the nature of Y_k. Hence for each aspect (criterion) a specific preference function must be defined. The most common preference functions are the following:

- *subjective*: values $+1$, 0 and -1 are assigned according to judgements of experts;
- *dichotomic*: -1 is assigned if a requested characteristic or property is satisfied by v but not by u, 0 is assigned if both units or neither of them satisfy the characteristic/property and $+1$ is assigned otherwise;
- *ordinal*: the k-th preference function takes the value $+1$ if $Y_{uk} > Y_{vk}$, -1 if $Y_{uk} < Y_{vk}$ and 0 otherwise;
- *ε-ordinal*: the k-th preference function takes the value $+1$ if $Y_{uk} > Y_{vk} + \varepsilon$, -1 if $Y_{uk} < Y_{vk} + \varepsilon$ and 0 otherwise;
- *α-stochastic*: value $+1$ (-1) is given if the observed value of Y_{uk} is greater (less) than the observed value of Y_{vk} and if they are stochastically not equal (at significance level α); value 0 is given otherwise.

Hence, considering Y_k, for each unit a flow is computed according to:

$$T_k(Y_{ck}) = \Phi_c^{(k)} = \sum_{v=1}^{C} h_k(Y_{ck}, Y_{vk}).$$

The flow measure the degree of preference associated to each unit. A positive flow express how much the unit dominates the other ones and a negative flow indicates how much it is dominated by the other ones. Based on these flows, K partial rankings of the C units are obtained. The global synthesis respect to the K aspects can be obtained through a weighted mean of flows (or of flow transformations which do not modify partial rankings) according to (10.7). For a review of the main multi-criteria methods related to the construction of a composite index see Gori and Vittadini (1999).

10.3 Global performance score

In general, in order to develop a new industrial product, a comparative evaluation of the performances of the new product with other existing ones is required. The performance has to be evaluated from several points of view and in different experimental conditions. Hence experimental data present a multidimensional structure and a stratification or blocking factor has to be considered in the design of the experiments and in the statistical data analysis. For example a chemical company operating in the field of detergents should be interested in comparative evaluations of detergents about cleaning efficacy, whiteness degree, etc. Each of these aspects can be represented by one variable. One or more variables represent primary performances and other variables represent secondary performances. To identify the best detergents, different kinds of stains, different kinds of textile, different numbers of washing cycles, etc. should be included in the experimental design. For this reason this is a typical example of a multistrata problem.

Without loss of generality, let us suppose that the primary performance is represented by Y_1 and secondary performances are represented by $Y_2, Y_3, \ldots, Y_K$. Moreover, let us consider two stratification factors and let us indicate with Y_{ctsk} the variable that represents the performance of product c, when the first stratum is at level t and the second is at level s, considering the k-th partial aspect (component response variable) with $c = 1, \ldots, C$, $t = 1, \ldots, T$, $s = 1, \ldots, S_t$, $\sum_{t=1}^{T} S_t = S$ and $k = 1, \ldots, K$. In this work we take into account a peculiar (but realistic) design where n experimental replications are performed to evaluate the primary performance but just one replication is available for the secondary performances. Hence the method proposed presents an initial step with two distinct procedures: the first one is dedicated to the calculation of an index for the primary performance, called Global Score on Primary performance (GSP); the second one is aimed to construct an index for the secondary performances, called Global Score on Secondary performance (GSS). The final step consists in the aggregation of GSP and GSS to obtain a global performance measure named Global Performance Score (GPS).

10.3.1 Global score on primary performance

Within each stratum an ANOVA test is performed and from the usual $C \times (C-1)/2$ pairwise comparisons it is possible to test the statistical significance of the differences between the mean performances for each couple of products (u, v). Let us indicate with $\bar{y}_{(1)ts1} \geq \bar{y}_{(2)ts1} \geq \ldots \bar{y}_{(C)ts1}$ the ordered observed sample means for Y_1 and assume that high values correspond to better performance. The algorithm to calculate the GSP is the following:

1. for each of the $T \times S$ strata a $C \times C$ matrix X is created (see the example in Table 10.1) where the elements under the main diagonal are null and those over the main diagonal take value 0 or 1 according to the following rule:

$$X[u,v] = h\left[\bar{y}_{(u)ts1}, \bar{y}_{(v)ts1}\right] \begin{cases} 1 & \text{if } \bar{y}_{(u)ts1} \text{ is significantly not equal to } \bar{y}_{(v)ts1} \\ 0 & \text{otherwise;} \end{cases}$$

2. a rank table, as shown in the example of Table 10.1, is created according to the following steps:

   a. in row 1, rank 1 is assigned to the product with the higher mean (first column), indicated with (1), and to all the other products whose mean performances are not significantly different from that of (1);
   b. in row 2, rank 2 is assigned to the product with the higher mean, among those excluded from rank 1 assignment, and to all the other products whose mean performances are not significantly different from that of (2);
   c. in row r, rank r is assigned to the product with the higher mean, among those excluded from rank $(r-1)$ assignment, and to all the other products whose mean performances are not significantly different from that of (r);
   d. the iterated procedure stops when a rank is assigned to the product (C);

3. for each product, the arithmetic mean of the values from the rank table (mean by columns) gives a partial performance score: $Z_{(c)ts1}$;
4. aggregated values $Z_{ct\cdot 1}$ respect to one stratification factor are obtained. For example if an additive aggregation is used,

$$Z_{ct\cdot 1} = \sum_{s=1}^{S_t} \pi_{ts} Z_{cts1};$$

while the application of a geometric rule gives

$$Z_{ct\cdot 1} = \prod_{s=1}^{S_t} Z_{cts1}^{\pi_{ts}}.$$

In general

$$Z_{ct\cdot 1} = g(Z_{ct11}, \ldots, Z_{ctS_t1}; \pi_{t1}, \ldots, \pi_{tS_t});$$

5. to have comparability among indicators and to facilitate the reading of the results, standardization has to be performed:

$$\tilde{Z}_{ct\cdot 1} = T_1(Z_{ct\cdot 1});$$

6. aggregation respect to the other stratification factor allows us to obtain a non-standardized measure of the primary performance for product c:

$$Z_{c\cdot\cdot 1} = g(\tilde{Z}_{c1\cdot 1}, \ldots, \tilde{Z}_{cT\cdot 1}; \pi_1, \ldots, \pi_T);$$

7. final standardization gives the requested Global Score on Primary performance:

$$GSP_c = \tilde{Z}_{c\cdot\cdot 1} = T_1(Z_{c\cdot\cdot 1}),$$

where $\pi_{11}, \ldots, \pi_{TS_T}, \pi_1, \ldots, \pi_T$ are predetermined weights.

10 A novel global performance score

Table 10.1 Example of X matrix for the multiple comparisons between pairs of products ($C=8$)

Ordered products	(1)	(2)	(3)	(4)	(5)	(6)	(7)	(8)
(1)		0	0	1	1	1	1	1
(2)			0	0	0	0	1	1
(3)				0	0	0	0	1
(4)					0	0	0	0
(5)						0	0	0
(6)							0	0
(7)								0
Rank								
1	1	1	1					
2		2	2	2	2	2		
3			3	3	3	3	3	
4				4	4	4	4	4
$Z_{(c)tsk}$	1	1.5	2	3	3	3	3.5	4

10.3.2 Global score on secondary performance

For variables $Y_2, Y_3, \ldots, Y_K$ just one replication is available, hence a synthetic measure of the secondary performances can be obtained by means of a sequential procedure of standardizations and aggregations. In general the secondary performance of product c, for a given t and a given s, related to the k-th aspect is calculated as a non decreasing function of the absolute difference $|Y_{ctsk} - Y^*_{tsk}|$ $(k = 2, 3, \ldots, K)$. Y^*_{tsk} is the value corresponding to the best theoretical or observed performance. Hence the starting point of the sequential procedure consists in a transformation which takes into account such value and allows standardization useful for the subsequent aggregation. The phases of the procedure are:

1. initial data transformation: $Z_{ctsk} = T_k \left(|Y_{ctsk} - Y^*_{tsk}| / CF_k \right)$, where CF_k represents the so-called "calibration factor", that is the a priori maximum difference supposed to exist between products with equal washing performance;
2. aggregation respect to one stratification factor:
   $Z_{ct \cdot k} = g(Z_{ct1k}, \ldots, Z_{ctS_tk}; \pi_{t1}, \ldots, \pi_{tS_t})$;
3. standardization: $\widetilde{Z}_{ct \cdot k} = T_k(Z_{ct \cdot k})$;
4. aggregation respect to the other stratification factor:
   $Z_{c \cdot \cdot k} = g(\widetilde{Z}_{c1 \cdot k}, \ldots, \widetilde{Z}_{cT \cdot k}; \pi_1, \ldots, \pi_T)$;
5. standardization: $\widetilde{Z}_{c \cdot \cdot k} = T_k(Z_{c \cdot \cdot k})$;
6. aggregation respect to the variables: $Z_{c \cdots} = g(\widetilde{Z}_{c \cdot \cdot 2}, \ldots, \widetilde{Z}_{c \cdot \cdot K}; \omega_2, \ldots, \omega_K)$;
7. final standardization: $GSS_c = \widetilde{Z}_{c \cdots} = T(Z_{c \cdots})$,

where $\pi_{11}, \ldots, \pi_{TS_T}, \pi_1, \ldots, \pi_T, \omega_2, \ldots, \omega_K$ are predetermined weights.

10.3.3 Aggregation of GSP and GSS

The final aggregation of GSP and GSS can be obtained using one of the considered aggregation technique. It is possible to use different aggregation methods at different levels of the procedure but, as suggested by Vitali and Merlini (1999), if it is possible and consistent with the specific application problem and with data, it is preferable to use the same aggregation function and standardization technique during all the steps of the procedure. Hence the final aggregation can be written as:

$$GPS_c = T[g(GSP_c, GSS_c; \tau_1, \tau_2)],$$

where τ_1 and τ_2 represents the degrees of importance assigned to primary performance evaluation and secondary performance evaluation respectively.

10.4 Case study: comparative performance evaluations of new detergents

As a real case study of GPS methodology, we propose to apply it to evaluate the performance of laundry detergents. In this context, the primary performace can be viewed as primary detergency, i.e. the assessment of the stain removal (cleaning) performance of a detergent (A.I.S.E., 2009). As secondary performance, we can refer to the so-called secondary detergency, that is the assessment of benefits which are measurable only after a certain minimum number of washing cycles (usually 5, 10 and 15). There are several useful secondary detergency performances: Whiteness Degree, Greying or Y-Value, Tint Value, Dye Fading and Dye Transfer Inhibition (A.I.S.E., 2009).

In the field of detergency, the GPS methodology can be applied to several types of common protocols, e.g.:

- cleaning stains in washing machines or laboratory scale equipments as a Lini-test and a Tergotometer;
- measuring performance results instrumentally or through expert panels;
- stains can be standard ones (i.e. EMPA167, CFT BC3, WFK 10J, etc.) or freshly prepared (i.e. ASTM D4265, Hohenstein, etc.).

Moreover, the GPS can be applied only to primary or secondary detergency or only to some parts of them, because to each variable/stratum corresponds a "weight", which can be set equal to zero if needed.

Let us consider the measuring of primary performance results through expert panels, hence we refer to a 5-point index where 1 means "stain completely removed" and 5 "stain completely not removed". Experimental replications are here provided by a visual evaluation of four experts (panelists). We consider the comparison of eight products and, as strata, three kinds of textile and several kinds of stains. Secondary performances are measured through instrumental values and are represented by three variables: Whitness Ganz (W), Graying (G) and Tint Value (T). Experiments are performed considering three different numbers of washing cycles (5, 10, 15) and several kinds of textile. Since the "target" values for the three secondary performances are the observed maximum for W and G and 0 for T, hence the initial transformations of data are:

- $Z_{cts}^{(W)} = \frac{\max_c(W_{cts}) - W_{cts}}{7}$;
- $Z_{cts}^{(G)} = \frac{\max_c(G_{cts}) - G_{cts}}{0.7}$;
- $Z_{cts}^{(T)} = \frac{|0 - T_{cts}|}{0.5} = \frac{|T_{cts}|}{0.5}$;

where denominators are the so called "calibration factors", i.e. scale parameters used in processing data calculated by means of previous secondary performance studies with many experimental replicates, the results of which have been statistically processed. The calibration factors are defined as the maximum difference reported between the experimental data relative to products with equal washing performance.

The weights of partial indexes, strata and variables have been set up by detergency experts of Reckitt-Benckiser R&D Division and are shown in Table 10.2. For primary performance, within each kind of textile, weights of stains are equal. For secondary performance, within each kind of textile, weights of washing cycles are the same.

GPS methodology has been applied to evaluate the global performance of the eight products. Different combinations of standardization techniques and aggregating functions are considered. The results are illustrated in Table 10.3. Methods (V) and (VI) differ from methods (III) and (IV) because are based on the theoretical maximum instead of the observed one.

As expected, different combinations of standardization and aggregation perform scores which range in a different domain but, of course, they keep the same ordering. In fact, the rankings in Table 10.3 are the same: B is always the best product whereas A and G are always the worst. The methods (VIII), (IX), (XII) and (XIII), with standardizations based on decreasing transformations (inverse and rescaling respect to the maximum), do not give results substantially different from the other methods but just a different reading key for the observed values. On the other hand methods (VIII) and (IX) allow to obtain values in $(0, 1)$ but cause a large reduction of the score variability. Moreover the inverse transformation does not remove the influence of the units of measurement of variables. If the goal of the analysis is the comparison of each product just with the best (or with the worst) product, methods (I), (II), (III) and/or (IV) are preferable. However methods (X), (XI), (XII) and (XIII) are the most useful, from the interpretation point of view, because extreme (best and worst) performances are indicated with values 0 and 1 (or viceversa) and for all the other products it is possible to see how much they are distant from the best but also from the worst performance. Methods (VII), (VIII) and (IX) give absolute evaluations, in the sense that the domain of the global index does not present a finite limit, but they are less useful from the interpretation point of view.

Table 10.2 Weights of partial indexes, textiles (strata) and variables in washing performance evaluations

Index/Variable/textile		weights
GSP (Global Score on Primary performance)		0.70
Cotton	0.70	
Polyester/Cotton	0.18	
Polyester	0.12	
GSS (Global Score on Secondary performance)		0.30
W (Whiteness)		0.33
Cotton	0.30	
Terry towel	0.25	
Single jersey	0.20	
Polyester/Cotton	0.15	
Polyamide	0.15	
G (Graying)		0.33
Cotton	0.25	
Terry towel	0.25	
Single jersey	0.15	
Polyester/Cotton	0.15	
Polyester	0.10	
Polyamide	0.10	
T (Tint Value)		0.33
Cotton	0.35	
Terry towel	0.25	
Single jersey	0.15	
Polyester/Cotton	0.15	
Polyamide	0.10	

Table 10.3 Detergent scores by methods according to GPS methodology

Method	Standardization	Aggregation	B	D	C	E	F	H	A	G
(I)	$\tilde{Z} = \frac{Z}{\min(Z)}$	additive	1.00	1.02	1.04	1.07	1.08	1.08	1.13	1.15
(II)		geometric	1.00	1.03	1.04	1.06	1.08	1.08	1.13	1.15
(III)	$\tilde{Z} = \frac{Z}{\max(Z)}$	additive	0.87	0.90	0.91	0.93	0.94	0.94	0.98	0.99
(IV)		geometric	0.78	0.80	0.88	0.92	0.93	0.94	0.97	0.98
(V)		additive	0.14	0.14	0.14	0.14	0.15	0.14	0.15	0.15
(VI)		geometric	0.10	0.10	0.10	0.11	0.11	0.11	0.11	0.11
(VII)	$\tilde{Z} = \frac{Z - \bar{Z}}{S_Z}$	additive	−0.85	−0.38	−0.39	−0.15	0.12	0.17	0.60	0.87
(VIII)	$\tilde{Z} = \frac{1}{Z}$	additive	0.12	0.12	0.12	0.11	0.11	0.11	0.11	0.11
(IX)		geometric	0.10	0.09	0.09	0.08	0.08	0.08	0.08	0.08
(X)	$\tilde{Z} = \frac{Z - \min(Z)}{\max(Z) - \min(Z)}$	additive	0.11	0.22	0.33	0.46	0.57	0.48	0.80	0.86
(XI)		geometric	0.00	0.09	0.21	0.32	0.31	0.20	0.80	0.80
(XII)	$\tilde{Z} = \frac{\max(Z) - Z}{\max(Z) - \min(Z)}$	additive	0.89	0.78	0.67	0.54	0.43	0.52	0.20	0.14
(XIII)		geometric	0.30	0.42	0.92	0.36	0.06	0.26	0.10	0.01

10.5 A comparative simulation study

In order to evaluate the degree of accuracy of GPS method in detecting the true "unknown" product global performance, in this section we perform a comparative simulation study with the goal of comparing the statistical performance of the set of all possible combinations of standardization and aggregation techniques. We underline that the proposed comparative simulation study has been developed with reference to the industrial experiment presented in the previous section, hence it mimics in details a real comparison among several existing products. Let us consider the following setting:

- 8 units (products) to be compared ($C = 8$), labelled A,B,...,H, with "hypothetical" true global performance as follows: $B = D > E = F > C = H > A = G$;
- for the primary performance, let 3 be the number of main strata ($T = 3$) and let $S_1 = 20$ and $S_2 = S_3 = 5$ be the number of levels for the second stratification factor;
- let 3 be the number of secondary performance variables ($K = 4$), and 5 be the number of main strata ($T = 5$) and $S_1 = \ldots = S_5 = 3$ the number of levels for the second stratification factor;
- the weights of strata and variables are shown in Table 10.2 (in previous section);
- we suppose to know, for each of the 8 products, the true partial performances; more specifically, we set the true partial performances as reported in Tables 10.4 and 10.5.

  Note that, since we have set the true partial performances we are also able to calculate the "true" GPS scores, for each combination of standardistation and aggregation techniques (Table 10.6). It is worth noting that the true "hypothetical" global performance (i.e. $B = D > E = F > C = H > A = G$) matches actually the 'true' product scores for each one of the thirteen considered GPS methods.

  We recall that primary performance are supposed to be generated by a replicated experimental design, hence Table 10.4 represents the true supposed mean values of each product by strata. Therefore, in order to perform our simulation study, we have to add to the true mean values an i.i.d. random component. For this goal, our simulation setting can be characterized by:

- a set of 1,000 random independent simulations, where for the primary simulated data we consider:

  - a fixed number of experimental replicates equal to 4;
  - two possible values for the variance of the random component: $\sigma_{cts} = 0.25$ or 0.5, $c = 1,\ldots,8$, $t = 1,\ldots,3$, $s = 1,\ldots,S_t$ ($S_1 = 20$ and $S_2 = S_3 = 5$);
  - three type of random errors: normal errors, exponential errors (as an example of an asymmetric distribution) and Student's t (with 2 d.f.; as an example of an heavy tailed distribution).

- With the aim of generating the secondary performaces, the values of Table 10.5 were added to a random error with variance equal to the half value of the corresponding calibration factor (i.e. $\sigma_W = 3.5$, $\sigma_G = 0.35$, $\sigma_T = 0.25$).

Table 10.4 "True" primary performances by product and strata

Str. fact. (Textile)	Str. fact. (Stain)	Product							
		A	B	C	D	E	F	G	H
Cotton	Tea	4	4	4	4	4	4	4	4
Cotton	Red wine	4.5	4.5	4.5	4.5	4	4	4.5	4.5
Cotton	Blueberry juice	3.5	3.5	3	3.5	3	3	3.5	3
Cotton	Cherry juice	2	2	2	2	2	2	2	2
Cotton	Chocolate dessert	2	2	2.5	2	2	2	2	2.5
Cotton	Cacao	3	2.5	3	2.5	2.5	2.5	3	3
Cotton	Chocolate ice cream	3	2.5	3	2.5	3	3	3	3
Cotton	Spinach	3	2.5	2.5	2.5	2.5	2.5	3	2.5
Cotton	Grass	3.5	3	3	3	3.5	3.5	3.5	3
Cotton	Carrots	3	3	4	3	4	4	3	4
Cotton	Chocolate pudding	1.5	1.5	1.5	1.5	1.5	1.5	1.5	1.5
Cotton	Curry sauce	4	3	3.5	3	3.5	3.5	4	3.5
Cotton	Tomato sauce	4	3.5	4	3.5	4	4	4	4
Cotton	Gravy	2	2	2	2	2	2	2	2
Cotton	Frying fat	2.5	2.5	2	2.5	2	2	2.5	2
Cotton	Lard	2.5	2	2	2	2	2	2.5	2
Cotton	Motor oil, mixture	5	5	4.5	5	4.5	4.5	5	4.5
Cotton	Make-up	5	5	4.5	5	4.5	4.5	5	4.5
Cotton	Lipstick	2.5	2.5	2.5	2.5	3	3	2.5	2.5
Cotton	Garden soil	4.5	4.5	4.5	4.5	4.5	4.5	4.5	4.5
Pol./Cot.	Frying fat	3.5	3	3.5	3	3.5	3.5	3.5	3.5
Pol./Cot.	Lard	2.5	2.5	2.5	2.5	2.5	2.5	2.5	2.5
Pol./Cot.	Motor oil	5	5	5	5	5	5	5	5
Pol./Cot.	Make-up	5	5	5	5	5	5	5	5
Pol./Cot.	Lipstick	3	2.5	2	2.5	2.5	2.5	3	2
Polyester	Frying fat	2.5	2	2	2	2	2	2.5	2
Polyester	Lard	1.5	1.5	1.5	1.5	1.5	1.5	1.5	1.5
Polyester	Motor oil	4	3.5	3.5	3.5	3.5	3.5	4	3.5
Polyester	Make-up	5	5	5	5	5	5	5	5
Polyester	Lipstick	2.5	2.5	2	2.5	2	2	2.5	2

In order to evaluate the degree of accuracy of all possible combinations of standardistation and aggregation methods, we consider the following three criteria:

1. the mean value of the Spearman's rank correlation coefficient (Spearman's ρ) between the performed GPS scores and the corresponding "true" GPS scores;
2. the mean square error (MSE) of the whole set of performed GPS scores (with reference to the "true" GPS scores);
3. the 95% "pseudo" confidence intervals for the GPS score of three selected products (B, F and G), that is the 0.025 and 0.975 percentiles of the empirical distribution we obtained from the 1,000 simulations for the GPS score. Note that the upper and the lower "pseudo" confidence limits represent the range where we observed the 95% of the individual GPS score distribution.

Results of the proposed simulation study are presented in Tables 10.7 and 10.8.

10 A novel global performance score

Table 10.5 "True" secondary performances by product and strata

Second. perform.	1. Str. fact. (Textile)	2. Str. fact. (Cycle)	Product							
			A	B	C	D	E	F	G	H
W	Cotton	5	187	195	190	195	191	191	187	190
W	Cotton	10	202	210	205	210	206	206	202	205
W	Cotton	15	212	220	215	220	216	216	212	215
W	Terry towel	5	222	230	225	230	226	226	222	225
W	Terry towel	10	232	240	235	240	236	236	232	235
W	Terry towel	15	237	245	240	245	241	241	237	240
W	Single jersey	5	222	230	225	230	226	226	222	225
W	Single jersey	10	227	235	230	235	231	231	227	230
W	Single jersey	15	229	237	232	237	233	233	229	232
W	Pol./Cot.	5	212	220	215	220	216	216	212	215
W	Pol./Cot.	10	214	222	217	222	218	218	214	217
W	Pol./Cot.	15	217	225	220	225	221	221	217	220
W	Polyamide	5	107	115	110	115	111	111	107	110
W	Polyamide	10	109	117	112	117	113	113	109	112
W	Polyamide	15	112	120	115	120	116	116	112	115
G	Cotton	5	90	91	90.5	91	90.5	90.5	90	90.5
G	Cotton	10	90	91	90.5	91	90.5	90.5	90	90.5
G	Cotton	15	90	91	90.5	91	90.5	90.5	90	90.5
G	Terry towel	5	90	91	90.5	91	90.5	90.5	90	90.5
G	Terry towel	10	89	90	89.5	90	89.5	89.5	89	89.5
G	Terry towel	15	89	90	89.5	90	89.5	89.5	89	89.5
G	Single jersey	5	89	90	89.5	90	89.5	89.5	89	89.5
G	Single jersey	10	89	90	89.5	90	89.5	89.5	89	89.5
G	Single jersey	15	89	90	89.5	90	89.5	89.5	89	89.5
G	Pol./Cot.	5	86	87	86.5	87	86.5	86.5	86	86.5
G	Pol./Cot.	10	86	87	86.5	87	86.5	86.5	86	86.5
G	Pol./Cot.	15	85	86	85.5	86	85.5	85.5	85	85.5
G	Polyester	5	86	87	86.5	87	86.5	86.5	86	86.5
G	Polyester	10	85	86	85.5	86	85.5	85.5	85	85.5
G	Polyester	15	85	86	85.5	86	85.5	85.5	85	85.5
G	Polyamide	5	89	90	89.5	90	89.5	89.5	89	89.5
G	Polyamide	10	87	88	87.5	88	87.5	87.5	87	87.5
G	Polyamide	15	86	87	86.5	87	86.5	86.5	86	86.5
T	Cotton	5	−1	−0.4	−0.7	−0.4	−0.6	−0.6	−1	−0.7
T	Cotton	10	−0.8	−0.2	−0.5	−0.2	−0.4	−0.4	−0.8	−0.5
T	Cotton	15	−0.6	0	−0.3	0	−0.2	−0.2	−0.6	−0.3
T	Terry towel	5	1.6	1	1.3	1	1.2	1.2	1.6	1.3
T	Terry towel	10	1.7	1.1	1.4	1.1	1.3	1.3	1.7	1.4
T	Terry towel	15	2.1	1.5	1.8	1.5	1.7	1.7	2.1	1.8
T	Single jersey	5	1.3	0.7	1	0.7	0.9	0.9	1.3	1
T	Single jersey	10	1.4	0.8	1.1	0.8	1	1	1.4	1.1
T	Single jersey	15	1.5	0.9	1.2	0.9	1.1	1.1	1.5	1.2
T	Pol./Cot.	5	1.6	1	1.3	1	1.2	1.2	1.6	1.3
T	Pol./Cot.	10	1.7	1.1	1.4	1.1	1.3	1.3	1.7	1.4
T	Pol./Cot.	15	2.1	1.5	1.8	1.5	1.7	1.7	2.1	1.8
T	Polyamide	5	0.9	0.3	0.6	0.3	0.5	0.5	0.9	0.6
T	Polyamide	10	1.2	0.6	0.9	0.6	0.8	0.8	1.2	0.9
T	Polyamide	15	1.6	1	1.3	1	1.2	1.2	1.6	1.3

Table 10.6 "True" product scores by methods according to GPS methodology

Method	Standardization	Aggregation	B	D	E	F	C	H	A	G
(I)	$\tilde{Z} = \frac{Z}{\min(Z)}$	additive	1.00	1.00	1.17	1.17	1.19	1.19	1.54	1.54
(II)		geometric	1.00	1.00	1.15	1.15	1.17	1.17	1.50	1.50
(III)	$\tilde{Z} = \frac{Z}{\max(Z)}$	additive	0.68	0.68	0.77	0.77	0.78	0.78	1.00	1.00
(IV)		geometric	0.44	0.44	0.64	0.64	0.67	0.67	1.00	1.00
(V)		additive	0.13	0.13	0.14	0.14	0.14	0.14	0.18	0.18
(VI)		geometric	0.09	0.09	0.11	0.11	0.11	0.11	0.14	0.14
(VII)	$\tilde{Z} = \frac{Z-\bar{Z}}{S_Z}$	additive	−0.90	−0.90	−0.33	−0.33	−0.25	−0.25	1.47	1.47
(VIII)	$\tilde{Z} = \frac{1}{Z}$	additive	0.14	0.14	0.11	0.11	0.11	0.11	0.08	0.08
(IX)		geometric	0.11	0.11	0.08	0.08	0.07	0.07	0.05	0.05
(X)	$\tilde{Z} = \frac{Z-\min(Z)}{\max(Z)-\min(Z)}$	additive	0.00	0.00	0.23	0.23	0.26	0.26	1.00	1.00
(XI)		geometric	0.00	0.00	0.12	0.12	0.07	0.07	1.00	1.00
(XII)	$\tilde{Z} = \frac{\max(Z)-Z}{\max(Z)-\min(Z)}$	additive	1.00	1.00	0.77	0.77	0.74	0.74	0.00	0.00
(XIII)		geometric	1.00	1.00	0.80	0.80	0.77	0.77	0.00	0.00

It is interesting to highlight that the considered thirteen methods (combinations of standardization and aggregation techniques) do not perform in the same way. The best one is the method labelled as IV, i.e. standardization by the maximum observed value and aggregation by multiplicative function. In fact, method IV is denoted by higher Spearman's ρ, lower MSE and lower width of the "pseudo" confidence interval. With reference to the last criterion, which seems to be the more consistent, note that method IV is the only one able to not overlay the "pseudo" confidence intervals of products B and E. In general, it is worth noting that not all methods have the same behaviour when the experimental variability is increasing and when the random distribution is changing (especially for Student's t error). Hence, some methods are more robust than others, in particular the most robust are VIII and IX and the less robust are the method labelled as VII and X-XIII.

10.6 Conclusions

The new method proposed in this chapter, named "Global Performance Score" or simply GPS, offers a suitable and consistent approach for the construction of a global performance index when we are facing a complex problem of product quality evaluation. As shown by the real case study and by the simulation study, GPS offers an immediate understanding of the results, thus facilitating their interpretation. The GPS methodology is based on data standardization and on aggregating techniques. Data standardization is necessary to compare different kinds of phenomena which have been included in the same performance evaluation: usually, these variables can be grouped into the so called "Primary" and "Secondary" performances. The GPS can be applied entirely to Primary and Secondary performances or just to some

Table 10.7 Simulation study results: mean value of the Spearman's rank correlation and MSE score

Method	Normal errors Spear.'s ρ	MSE	Exponential errors Spear.'s ρ	MSE	Student's t errors Spear.'s ρ	MSE
			$\sigma=0.25$			
(I)	0.89	0.081	0.89	0.090	0.87	0.263
(II)	0.89	0.078	0.88	0.084	0.88	0.265
(III)	0.88	0.068	0.88	0.072	0.87	0.282
(IV)	0.92	0.061	0.92	0.071	0.91	0.180
(V)	0.82	0.072	0.81	0.075	0.83	0.099
(VI)	0.89	0.099	0.88	0.101	0.88	0.080
(VII)	0.84	0.154	0.84	0.150	0.71	0.412
(VIII)	0.87	0.067	0.87	0.073	0.88	0.093
(IX)	0.92	0.070	0.92	0.081	0.91	0.085
(X)	0.84	0.115	0.84	0.111	0.72	0.286
(XI)	0.79	0.114	0.80	0.113	0.76	0.262
(XII)	0.84	0.115	0.84	0.111	0.72	0.286
(XIII)	0.81	0.140	0.80	0.138	0.71	0.350
			$\sigma=0.5$			
(I)	0.91	0.195	0.90	0.211	0.87	0.340
(II)	0.90	0.219	0.90	0.229	0.89	0.362
(III)	0.89	0.254	0.89	0.253	0.88	0.390
(IV)	0.93	0.138	0.92	0.152	0.91	0.212
(V)	0.88	0.121	0.87	0.111	0.83	0.175
(VI)	0.90	0.060	0.90	0.061	0.89	0.087
(VII)	0.74	0.400	0.75	0.384	0.55	0.589
(VIII)	0.90	0.047	0.89	0.053	0.90	0.088
(IX)	0.93	0.035	0.92	0.045	0.91	0.071
(X)	0.72	0.304	0.73	0.290	0.55	0.382
(XI)	0.78	0.283	0.79	0.261	0.62	0.362
(XII)	0.72	0.304	0.73	0.290	0.55	0.382
(XIII)	0.69	0.328	0.70	0.317	0.61	0.398

parts of them: as a matter of fact to each partial aspect of performance corresponds a "weight", which can be set to zero if needed.

When considering a global performance index it is possible the application of different standardization methods and aggregating techniques. In the application problem described in this chapter, in general, the rankings coming from different combinations of standardization and aggregations are obviously similar but they can substantially differ in accuracy. Moreover, the methods with standardizations based on decreasing transformations (inverse and rescaling respect to the maximum) do not give results substantially different from the other methods but just a different reading key for the observed values. Methods based on inverse transformation allow to obtain values in $(0,1)$ but cause a large reduction of the variability and of performance differences between products and the influence of the units of mea-

Table 10.8 Simulation study results: "pseudo" confidence intervals (with $\sigma = 0.25$)

Method	product B lower lim.	product B upper lim.	product E lower lim.	product E upper lim.	product C lower lim.	product C upper lim.	product A lower lim.	product A upper lim.
			Normal	errors				
(I)	1.00	1.12	1.10	1.27	1.12	1.28	1.39	1.58
(II)	1.00	1.11	1.09	1.25	1.10	1.26	1.35	1.55
(III)	0.67	0.78	0.73	0.86	0.74	0.86	0.95	1.00
(IV)	0.44	0.55	0.59	0.73	0.61	0.76	0.92	1.00
(V)	0.14	0.16	0.14	0.17	0.15	0.17	0.18	0.21
(VI)	0.11	0.12	0.12	0.14	0.12	0.14	0.15	0.17
(VII)	−1.22	−0.31	−0.73	0.15	−0.66	0.23	0.93	1.59
(VIII)	0.11	0.12	0.09	0.10	0.09	0.10	0.07	0.08
(IX)	0.08	0.10	0.06	0.07	0.06	0.07	0.04	0.05
(X)	0.00	0.29	0.12	0.53	0.14	0.55	0.78	1.00
(XI)	0.00	0.11	0.00	0.44	0.01	0.46	0.78	1.00
(XII)	0.71	1.00	0.47	0.88	0.45	0.86	0.00	0.22
(XIII)	0.53	1.00	0.30	0.92	0.39	0.89	0.00	0.06
			Expon.	errors				
(I)	1.00	1.12	1.10	1.25	1.10	1.26	1.37	1.56
(II)	1.00	1.11	1.08	1.23	1.09	1.24	1.34	1.53
(III)	0.67	0.78	0.74	0.85	0.74	0.87	0.95	1.00
(IV)	0.46	0.56	0.59	0.73	0.62	0.76	0.90	1.00
(V)	0.14	0.17	0.15	0.17	0.15	0.17	0.18	0.21
(VI)	0.11	0.13	0.12	0.14	0.12	0.14	0.15	0.17
(VII)	−1.19	−0.33	−0.70	0.12	−0.67	0.22	0.98	1.62
(VIII)	0.10	0.12	0.09	0.10	0.09	0.10	0.07	0.08
(IX)	0.07	0.09	0.06	0.07	0.06	0.07	0.04	0.05
(X)	0.00	0.29	0.11	0.50	0.14	0.55	0.79	1.00
(XI)	0.00	0.11	0.00	0.41	0.01	0.46	0.76	1.00
(XII)	0.71	1.00	0.50	0.89	0.45	0.86	0.00	0.21
(XIII)	0.50	1.00	0.30	0.91	0.40	0.89	0.00	0.06
			Stud.'s t	errors				
(I)	1.00	1.14	1.05	1.28	1.07	1.28	1.18	1.42
(II)	1.00	1.10	1.05	1.22	1.06	1.23	1.16	1.35
(III)	0.77	0.89	0.83	0.95	0.83	0.96	0.93	1.00
(IV)	0.50	0.67	0.64	0.86	0.67	0.89	0.86	1.00
(V)	0.12	0.14	0.13	0.15	0.13	0.16	0.14	0.17
(VI)	0.10	0.12	0.11	0.13	0.11	0.13	0.12	0.14
(VII)	−1.29	0.14	−0.78	0.66	−0.81	0.72	−0.16	1.40
(VIII)	0.10	0.12	0.09	0.11	0.09	0.11	0.08	0.09
(IX)	0.07	0.10	0.06	0.08	0.05	0.07	0.04	0.06
(X)	0.00	0.57	0.12	0.86	0.13	0.88	0.39	1.00
(XI)	0.00	0.22	0.00	0.72	0.01	0.76	0.02	1.00
(XII)	0.43	1.00	0.14	0.88	0.12	0.87	0.00	0.61
(XIII)	0.26	1.00	0.01	0.89	0.01	0.88	0.00	0.23

surement of variables is not removed. If the goal of the analysis is the comparison of each product just with the best (or just with the worst) product, standardizations respect to the observed maximum or to the observed minimum methods are preferable. However rescaling-based methods are the most useful, from the interpretation point of view, because for each product it is possible to measure the degree of proximity to the best but also to the worst performance. Methods based on inverse transformation and on standardization give absolute evaluations, in the sense that the domain of the global index does not present a finite limit, but they are less useful from the interpretation point of view.

Chapter 11
Nonparametric tests for the randomized complete block design with ordered categorical variables

Livio Corain and Luigi Salmaso

11.1 Introduction

In many scientific disciplines and industrial fields, when dealing with comparisons between two or more treatments, researchers and practitioners are often faced with theoretical and practical problems within the framework of Randomized Complete Block (RCB) design with ordered categorical response variables. This situations can arise very often in the field of the evaluation of educational services or quality of products, for example in connection with the sensorial testing studies, where several useful experimental performance indicators, especially in the food and body care industry, are provided by individual sensorial evaluations by trained people (panelists) during a so-called sensory test (Meilgaard et al., 2006). Within this framework the experimental design typically handles panelists as blocks.

In general, the requirement to take into consideration a RCB design occurs when the experimental units are heterogeneous, hence the notion of blocking is used to control the extraneous sources of variability. The major criteria of blocking are characteristics associated with the experimental material and the experimental setting. The purpose of blocking is to sort experimental units into blocks, so that the variation within a block is minimized while the variation among blocks is maximized. An effective blocking not only yields more precise results than an experimental design of comparable size without blocking, but also increases the range of validity of the experimental results.

In this contribution we propose a general solution within the Nonparametric Combination (NPC) of Dependent Permutation Tests (Pesarin, 2001) which is

---

Livio Corain
Department of Management and Engineering, University of Padua, Str. S. Nicola 4, 36100 Vicenza, Italy, e-mail: livio.corain@unipd.it

Luigi Salmaso
Department of Management and Engineering, University of Padua, Str. S. Nicola 4, 36100 Vicenza, Italy, e-mail: salmaso@gest.unipd.it

particularly suitable for the RCB design, especially in case of ordered categorical response variables such that used for sensorial studies. In the next section, we present an update review of the procedures proposed in the literature for the hypothesis testing on the RCD design. In Sect. 11.3 we present the proposed permutation solution for the RCB Design. In Sects. 11.4 and 11.5 a comparative simulation study and a real case study are presented. Finally, we conclude, in Section 6, with some directions of current and future research.

11.2 Overview on procedures proposed in the literature for the RCB design

Let us consider the experimental design where there are n blocks and, within each block, experimental units are randomly assigned to the C treatments ($C > 2$) and exactly one unit is assigned to each of the C treatments. The statistical model (with fixed effects) for the randomized complete block (RCB) design can be represented as follows:

$$Y_{ij} = \mu + \beta_i + \tau_j + \varepsilon_{ij}, \varepsilon_{ij} \sim IID(0, \sigma^2), i = 1,...,n, j = 1,...,C, \tag{11.1}$$

where β_i, τ_j and Y_{ij}, are respectively the effect of the i-th block, the effect of the j-th treatment and the response variable for the i-th block and the j-th treatment. The random term ε_{ij} represents the experimental error with zero mean, variance σ^2 and unknown continuous distribution P. The usual side-conditions for effects are given by the constrains $\sum_i \beta_i = \sum_j \tau_j = 0$.

Model (11.1) is called "effect model" (Montgomery, 2005). If we define $\mu_j = \mu + \tau_j$, $j = 1,...,C$, an alternative representation of model (11.1) is the so called "mean model", i.e.

$$Y_{ij} = \mu_j + \beta_i + \varepsilon_{ij}. \tag{11.2}$$

The resulting inferential problem of interest is concerned with the following hypotheses: $H_0 : \{\tau_j = 0, \forall j\}$, against $H_1 : \{\exists j : \tau_j \neq 0\}$. Note that this hypothesis is referred to a global test; if H_0 is rejected, it is of interest to perform inference on each pairwise comparison between couples of treatments, i.e. $H_{0(jh)} : \tau_j = \tau_h$, $j, h = 1,...,C$, $j \neq h$, against $H_{1(jh)} : \tau_j \neq \tau_h$; with reference to model (11.2), an equivalent representation of $H_{0(jh)}$ is the following: $H_{0(jh)} : \mu_j - \mu_h = 0$, $j, h = 1,...,C$, $j \neq h$, against $H_{1(jh)} : \mu_j - \mu_h \neq 0$.

We recall that in the framework of RCB designs there is usually no interest in testing the block effect which is handled as a nuisance factor. Note that, since no interaction effect between treatments and blocks is here supposed to exist, expressions (11.1) and (11.2) do not consider any interaction effect.

11 Nonparametric tests for the RCB design with ordered categorical variables

In the framework of traditional parametric methods, when assuming random normal components, it is appropriate to test the equality of all treatment means by using the traditional F statistic:

$$F = \frac{SS_{\text{Treatments}}/(C-1)}{SS_E/(n-1)(C-1)}, \qquad (11.3)$$

where $SS_{\text{Treatments}} = n \sum_{j=1}^{C}(\overline{Y}_{\cdot j} - \overline{Y}_{\cdot\cdot})^2$, $SS_E = \sum_{i=1}^{n}\sum_{j=1}^{C}(Y_{ij} - \overline{Y}_{\cdot j} - \overline{Y}_{i\cdot} + \overline{Y}_{\cdot\cdot})^2$ and $\overline{Y}_{\cdot j}$ is the mean of the n experimental units in the j-th treatment, $\overline{Y}_{i\cdot}$ is the block mean for the i-th block, and $\overline{Y}_{\cdot\cdot}$ is the overall mean. The F statistic is distributed as $F_{C-1,(C-1)(n-1)}$ if the null hypothesis H_0 is true, hence we would reject H_0, at the significance level α, if $F_0 > F_{\alpha;(C-1),(C-1)(n-1)}$. If the analysis indicates a significant difference in treatment means, we are usually interested in multiple comparisons to find out which treatment means differ. That is, when the global null hypothesis H_0 would be rejected we would consider the post-hoc set of $C(C-1)/2$ individual $H_{0(jh)}$ null hypotheses. Under normality, Bonferroni adjusted t-tests or Tukey's tests are the most recommended procedures. We recall that when carrying out multiple testing, there should be a formal guarantee against incorrect decisions. The so called multiplicity problem is particularly relevant in multiple comparison problems, since omitting to consider the multiplicity issue can often cause biased statistical analyses (Westfall et al., 1999).

Since the normality assumption is often questionable, if we do not assume the normality of random errors we can take into consideration a nonparametric approach. In the framework of nonparametric rank-based testing procedures, one of the earlier tests has been proposed by Friedman (1937). A general form of the Friedman's statistic T, which incorporates a correction for ties (Lehmann and D'Abrera, 2006), is given by:

$$T = \frac{(C-1)\sum_{j=1}^{C}[R_{+j} - n(C+1)/2]^2}{\sum_{i=1}^{n}\sum_{j=1}^{C}(R_{ij})^2 - nC(C+1)^2/4}, \qquad (11.4)$$

where R_{ij} is the rank of Y_{ij} among the experimental units in block i and $R_{+j} = \sum_j R_{ij}$ is the sum of the ranks for the j-th treatment over the n blocks. Under the null hypothesis, the R_{+j}'s should be close to $n(C+1)/2$ which is the average of the R_{+j}. Since T has an asymptotic Chi-square distribution with $C-1$ degree of freedom, we would reject the null hypothesis H_0 if $T_0 > \chi^2_{\alpha,C-1}$. After rejection of H_0, the comparisons between pairs of treatments can be performed via absolute differences of the sums of within-blocks ranks. This set of values have to be compared with an appropriate value r_α which is function of C and n. For small values of C and n, r_α has been tabulated whereas, as n tends to infinity, it can be approximated by the distribution of the range of independent standard normal variables. This procedure, called Wilcoxon-Nemenyi-McDonald-Thompson procedure (Hollander and Wolfe, 1999), has been designed in order to maintain an appropriate Maximum Experimentwise

Error Rate (MEER) α, where EER is defined as the probability to reject at least one true hypotheses in the set of $C(C-1)/2$ individual $H_{0(jh)}$ null hypotheses.

Following Lehmann and D'Abrera (2006), the formula (11.4) can be replaced by:

$$T = nd'\Sigma_0^{-1}d, \tag{11.5}$$

where $\Sigma_0 = (\sigma_{jj'})$ is the covariance matrix under the null hypothesis of $R_i = (R_{i1}, \cdots, R_{i,C-1})$, that is the rank order of the first $C-1$ treatments, and

$$d' = \left[R_{+1} - (C+1)/2, R_{+2} - (C+1)/2, \cdots, R_{+(C-1)} - (C+1)/2\right], \tag{11.6}$$

where $R_{+j} = \sum_j R_{ij}$. Sepansky (2007) suggests a modification of (11.6), by the following test statistic:

$$T_P = nd'\widehat{\Sigma}^{-1}d, \tag{11.7}$$

where $\widehat{\Sigma}^{-1} = (s_{jj'})$ is the sample covariance matrix of the R_i. Note that T_P is an Hotelling-type T^2 statistic and its limiting distribution is the χ^2 distribution with $C-1$ degrees of freedom (see Hollander and Wolfe, 2003). Sepansky (2007) examines also the covariance matrix in the test statistic (11.7) when the number of blocks or sample size is small and he claims that the null hypothesis of no treatment difference should be rejected when the sample covariance matrix is singular. It is worth noting that while the Friedman test statistic is well defined when n is less than C, T_P is not since the sample covariance matrix is singular for all possible data matrices in this case. The idea of Sepanky of rejecting the null hypothesis when the sample covariance matrix is questionable and he does not support this statement with any kind of formal proof and the motivation he provided is quite debatable. Moreover, the simulation results presented by author clearly show that, especially for small values of n, his test statistic does not maintain the nominal levels under the null hypothesis. Hence, this proposal might be unreliable to properly perform inference for RCB designs.

Another approach, refereed as aligned rank test (Lehmann and D'Abrera, 2006), is to make all blocks comparable so that comparisons between treatments in different blocks are meaningful. This can be done by subtracting the median or mean value of the experimental units in the block from all experimental units in that block. After this alignment is completed, the aligned experimental units are ranked over all blocks and treatments. It can be shown that, under the null hypothesis, the following statistic is a χ^2_{C-1} for large samples:

$$S = \frac{(C-1)n^2 \sum_{j=1}^{C} \left(\overline{R}_{\cdot j} - \overline{R}_{\cdot\cdot}\right)^2}{\sum_{i=1}^{n} \sum_{j=1}^{C} \left(R_{ij} - \overline{R}_{i\cdot}\right)^2}, \tag{11.8}$$

where now R_{ij} denotes the aligned rank for Y_{ij}, $\bar{R}_{i\cdot}$ is the average rank for the i-th block, $\bar{R}_{\cdot j}$ is the average rank for the j-th treatment and $\bar{R}_{\cdot\cdot}$ is the overall average rank.

In the literature there are a few other test statistics proposed for the RCB design. Among others, Quade (1979) proposed a test based on within-block rankings that gives greater weights to blocks that have greater variability. However, since several simulations studies (Fawcett and Salter, 1984); Groggel (1987) have shown that the Quade procedure is not well performing in some situations, hence as suggested by O'Gorman (2001), it will be not included in the simulations we will present afterwards in this work. O'Gorman (2001) reviews and evaluates several tests for RCB design, including the F-test, Friedman's test, and a few aligned rank tests. His simulations show that Friedman's test has low power compared with the aligned rank tests if the number of treatments does not exceed six and a novel aligned rank-based F-test proposed by the author shows relatively high power for several skewed distributions if there is a large number of experimental units.

11.3 Permutation tests for multivariate RCB design

When dealing with complex designs conditional nonparametric methods can represent a reasonable approach. We recall that traditional unconditional parametric testing methods (such as t test or F test) may be available, appropriate and effective only when a set of restrictive conditions are satisfied. Accordingly, just as there are circumstances in which unconditional parametric testing procedures may be appropriate, there are others where they may be unsuitable or even impossible to be properly applied. In conditional testing procedures, provided that exchangeability of data with respect to groups is satisfied in the null hypothesis, permutation methods play a central role. This is because they allow for quite efficient solutions, are useful when dealing with many difficult problems, provide clear interpretations of inferential results, and allow for weak extensions of conditional to unconditional inferences. For a detailed discussion on the topic of the comparison between permutation conditional inferences with traditional unconditional inferences we refer to Pesarin (2002).

In this chapter we propose a novel solution for the whole set of hypotheses of interest within the nonparametric framework of NonParametric Combination (NPC) of dependent permutation tests (Pesarin, 2001; Corain and Salmaso, 2004).

In order to better explain the proposed approach let us denote an $(n \times C)$ data set $\mathbf{Y}$ as:

$$\mathbf{Y} = \begin{bmatrix} \mathbf{Y}_1, ..., \mathbf{Y}_j, ..., \mathbf{Y}_C \end{bmatrix} = \begin{bmatrix} Y_{11} & ... & Y_{1j} & ... & Y_{1C} \\ ... & ... & ... & ... & ... \\ Y_{i1} & ... & Y_{ij} & ... & Y_{iC} \\ ... & ... & ... & ... & ... \\ Y_{n1} & ... & Y_{nj} & ... & Y_{nC} \end{bmatrix},$$

where Y_{ij} represents the ijth observed response for ith block and jth treatment, $i = 1,...,n$, $j = 1,...,C$, $(C \geq 2)$.

In the framework of NonParametric Combination (NPC) of dependent permutation tests we suppose that, if the global null hypothesis H_0 is true, the hypothesis of exchangeability of random errors within the same block holds. Hence, the following set of mild conditions should be jointly satisfied:

(i) Suppose that for $\mathbf{Y} = [\mathbf{Y}_1,...,\mathbf{Y}_C]$ an appropriate distribution P_j exists, $P_j \in \mathscr{F}$, $j = 1,...,C$, belonging to a (possibly non-specified) family $\mathscr{F}$ of non-degenerate probability distributions;

(ii) The null hypothesis H_0 states the equality in distribution of the response variable in all C groups:

$$H_0 = [P_1,...,P_C] = [\mathbf{Y}_1 \stackrel{d}{=} ... \stackrel{d}{=} \mathbf{Y}_C].$$

Null hypothesis H_0 implies the exchangeability, within each block, of the individual data with respect to the C groups. Moreover H_0 is supposed to be properly decomposed into $C \times (C-1)/2$ sub-hypotheses $H_{0(jh)}$, $j,h = 1,...,C$, $j \neq h$, each one related to the jhth pairwise comparison between couples of treatments:

$$H_0 = [\bigcap_{\substack{j,h=1 \\ j \neq h}}^{C} \mathbf{Y}_j \stackrel{d}{=} \mathbf{Y}_h] = [\bigcap_{\substack{j,h=1 \\ j \neq h}}^{C} H_{0(jh)}].$$

H_0 is called the *global* or *overall null hypothesis*, and $H_{0(jh)}$, $j,h = 1,...,C$, $j \neq h$, are the *partial null hypotheses*.

(iii) The alternative hypothesis H_1 is represented by the union of partial $H_{1(jh)}$ sub-alternatives:

$$H_1 = [\bigcup_{\substack{j,h=1 \\ j \neq h}}^{C} H_{1(jh)}] = [\bigcup_{\substack{j,h=1 \\ j \neq h}}^{C} H_{1(jh)}],$$

so that H_1 is true if at least one of sub-alternatives is true.

In this context, H_1 is called the *global* or *overall alternative*, and $H_{1(jh)}$, $j,h = 1,...,C$, $j \neq h$, are called the partial alternatives.

(iv) Let $\mathbf{T} = \mathbf{T}(\mathbf{Y})$ represent a vector of test statistics, whose components $T_{(jh)}$, $j,h = 1,...,C$, $j \neq h$, represent the partial univariate and non-degenerate *partial test* appropriate for testing the sub-hypothesis $H_{0(jh)}$ against $H_{1(jh)}$. Without loss of generality, all partial tests are assumed to be marginally unbiased, consistent and significant for large values (for more details see Pesarin, 2001).

At this point, in order to test the global null hypothesis H_0 and the $C \times (C-1)/2$ hypotheses $H_{0(jh)}$, we perform the partial (univariate) tests and then we combine them, with an appropriate combining function, in order to test the global null hypothesis H_0.

However, we should observe that in most real problems when the number of blocks is large enough, there might be computational difficulties in calculating the conditional permutation distribution. This means that it is not possible to calculate the exact p-value of observed statistic $T_{(jh)0}$. This drawback is overcome by using the Conditional Monte Carlo (CMC) Procedure. The CMC on the pooled data set **Y** is a random simulation of all possible permutations of the same data under H_0 (for more details refer to Pesarin, 2001). Hence, in order to obtain an estimate of the permutation distribution under H_0 of all test statistics, a CMC can be used. It should be emphasized that CMC only considers permutations of individual data vectors within each individual block, so that all underlying dependence relations which are present in the component variables are preserved. From this point of view, the CMC is essentially a multivariate procedure.

A suitable algorithm for calculating the proposed permutation test is composed of the following steps:

(a) For each pairwise comparison between couples of treatments calculate the vector of the observed values of test statistics $^oT(Y)$, whose components $^oT_{jh} = T(\mathbf{Y}_j, \mathbf{Y}_h)$, $j, h = 1, \ldots, C$, $j \neq h$, are appropriate for testing the sub-hypothesis $H_{0(jh)}$ against $H_{1(jh)}$.

(b) Consider $\mathbf{Y}^*$ as a permutation of the data set **Y**, carried out within each ith block in order to preserve the dependence structure of data, then calculate the permutation value of the test statistics:

$$T_{jh}^* = T\left(\mathbf{Y}_j^*, \mathbf{Y}_h^*\right), \quad j, h = 1, \ldots, C, \ j \neq h.$$

(c) Carry out B independent repetitions (i.e. Conditional Monte Carlo, CMC, iterations) of step (b). The set of CMC results $\left\{_bT_{jh}^*, \ b = 1, \ldots, B\right\}$ is thus a random sampling from the permutation distribution of the test statistics.

(d) Obtain the p-value from each partial sub-hypothesis $H_{0(jh)}$:

$$\lambda_{jh} = \#\left(T_{jh}^* \geq^o T_{jh}\right)/B, \quad b = 1, \ldots, B, \ j, h = 1, \ldots, C, \ j \neq h.$$

(e) The combined observed value of the global or overall null hypothesis H_0 is:

$$^oT'' = \psi(\lambda_{11}, \ldots, \lambda_{(C-1)C}).$$

(f) The combined value is then computed by:

$$T''^* = \psi\left(\lambda_{11}^*, \ldots, \lambda_{(C-1)C}^*\right).$$

where $\lambda_{jh}^* = \#\left(T_{jh}''^* \geq _bT_{jh}''^*\right)/B, \quad b = 1, \ldots, B.$

(g) The global p-value is computed as:

$$\lambda'' = \#(T''^* \geq {^oT''}), \quad b = 1, \ldots, B.$$

Matlab routines implementing permutation test for RCB design are available upon request by authors.

It can be seen that under the general null hypothesis the CMC procedure provides a consistent estimation of the permutation distributions, both marginal and combined, of the S partial tests. In the nonparametric combination procedure, Fisher's combination function is usually considered, principally for its good properties which are both finite and asymptotic (Pesarin, 2001). Of course, if it were considered appropriate, it would be possible to take into consideration any other combining function. The combined test is unbiased and consistent.

A general characterization of the class of combining functions is given by the following three main features for the combining function ψ:

(a) it must be non-increasing in each argument:

$$\psi(\ldots,\lambda_s,\ldots) \geq \psi(\ldots,\lambda'_s,\ldots) \text{ if } \lambda_s < \lambda'_s, s \in \{1,\ldots,S\};$$

(b) it must attain its supreme value, possibly not finite, even when only one argument reaches zero:

$$\psi(\ldots,\lambda_s,\ldots) \to \overline{\psi} \text{ if } \lambda_s \to 0, s \in \{1,\ldots,S\};$$

(c) $\forall \alpha > 0$, the critical value of every ψ is assumed to be finite and strictly smaller than the supreme value:

$$T''_\alpha < \overline{\psi}.$$

The above properties define the class C of combining functions. Some of the functions most often used to combine independent tests (Fisher, Lancaster, Liptak, Tippett, Mahalanobis, etc.) are included in this class. For a detailed description on how to build partial and global permutation tests refer to Pesarin (2001) and Corain and Salmaso (2004).

11.4 Simulation study

In order to validate the proposed method and to evaluate its performance in comparison with either the traditional parametric (F and t test) and the nonparametric approach (Friedman and aligned rank tests), in this section we perform a comparative simulation study. The goal is focused either on the global test H_0 and on the related treatment pairwise comparisons (hypotheses $H_{0(jh)}$).

The real context we are referring to is a typical sensorial study where the number of blocks (panel lists) usually ranges around 10–15 people and the sensorial evaluation is provided with a Likert 1–5 rating ordinal scale, where we suppose that the 0.5 scores are admitted as well. Note that we are actually considering a 9 point ordered categorical response variable.

Let us consider the following setting:

- 1,000 independent simulations;
- number of blocks: $n = 6, 10, 20$; number of treatment: $C = 3, 5, 7$;
- block effect β_i, $i = 1,\ldots,n$, generated from a discrete uniform distribution with values $(-1, -0.5, 0, 0.5, 1)$;

- with reference to model (11.2), the treatment effects μ_j, $j = 1,..,C$, are set in Fig. 11.1.
- three type of random errors: normal, exponential (as an example of an asymmetric distribution) and Student's t with 2 degree of freedom (as an example of an heavy tailed distribution). The variability of random errors has been calibrated to the value of $\sigma = 2$, with the aim of properly reveal and compare the power among the considered procedures. Finally, in order to better represent a genuine ordinal scale, before being added to the true effects the random errors were rounded to the nearest integer.

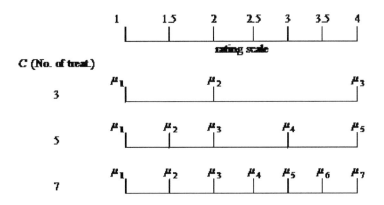

Fig. 11.1 Scheme of treatment effects for the simulation study.

For each simulation we performed the permutation tests (with 1,000 CMC), using the Fisher combining function, and we considered as counterparts the traditional F-test, the Friedman test and finally the Mean Aligned Rank (MAR) test proposed by O'Gorman (2001). The considered significance level was $\alpha = 0.05$. In case of rejection of the global null hypothesis H_{0k}, in order to perform the treatment pairwise comparisons, we considered permutation tests for two paired samples. Least Significant Difference (LSD) for the difference of mean ranks and t-tests as post-hoc procedures respectively for Friedman test and F-test and MAR have been considered as well. We recall that all post-hoc pairwise procedures should take into account for the problem of multiplicity (Westfall et al., 1999) hence they have to be well defined in order to maintain at the desired α-level the type I error probability of the main global hypothesis H_0. For this goal, for permutation tests we adopted a multiplicity correction strategy by using the closed testing approach (Marcus et al., 1976) via Tippett combining function (i.e. the so called minP procedure, Westfall et al., 1999) which is particularly suitable to be implemented within the framework of permutation tests (Finos and Salmaso, 2007), while for all other pairwise procedures we adopted the Bonferroni correction. Table 11.1 summarizes the obtained rejection rates ($\alpha = 0.05$). Note that, in order to be able to properly compare the performances of the compared procedures with different values of C (i.e. no. of treatments), rejection rates of pairwise comparisons are presented in terms of *delta* (δ), that is of the true differences (in term of σ) between treatment effects, where delta is defined as

$$\delta_{jh} = \tau_j - \tau_h, \; j,h = 1,...,C, \; j \neq h.$$

For example we get $\delta = 1\sigma$ for $C = 3$ from the difference between μ_2 and μ_1, whereas we get $\delta = 1\sigma$ for $C = 5$ from the differences $\mu_3 - \mu_1$, $\mu_4 - \mu_3$, $\mu_5 - \mu_4$.

As first remark for the simulation study, we can observe that under null hypothesis all procedures appear to properly behave according to the nominal level. From a general point of view, as expected, the power for the global hypothesis increases when increasing the number of blocks and the number of active treatments. On the contrary, power for pairwise comparisons decreases from 3 to 5 treatments and slightly increases from 5 to 7. This is probably due to a drawback of the multiplicity correction strategy which is too much conservative.

Obviously, F-test shows a better behaviour under normality, but in case of exponential errors and particularly of Student's t errors, all nonparametric procedures show a greater power. Among nonparametric tests, the worst one is the Friedman test whereas a good behaviour is provided by the Mean Aligned Rank test. It should be noted that Friedman test is actually not satisfactory when data have ties as in case of ordered categorical variables we considered in this chapter. In fact, the continuity correction proposed by several authors is valid only asymptotically and for finite samples it does not provides a conservative test. Permutation test has an intermediate performance which is denoted by some strength and weakness aspects: it is particularly powerful when the number of treatments is not too high and the number of blocks is around ten. An advantage of the permutation method is that it can be easily extended to the multivariate case, i.e. when the response variable in multidimensional, by means of the nonparametric combination methodology (Pesarin, 2001).

11.5 Case study

In this section we face a real case study proposed in the literature. Suppose, as in Lamond (1970), p. 28, that we wish to compare the flavour of meat from three breeds of geese X, Y, and Z on a five point scale with categories ranging from "excellent" to "very poor" and that the data from eight consumers shown in Table 11.2 are obtained, where we have labelled the ordered categories as 1–5 scores.

When applying the considered RCB procedures to meat flavour data we can obtain results reported in Table 11.3, where we performed pairwise comparisons only if the global test had been rejected ($\alpha = 0.05$). Note that, in addition to the Fisher combining function, we considered here for the global test Tippett and Liptak combining functions (Pesarin, 2001).

11 Nonparametric tests for the RCB design with ordered categorical variables

Table 11.1 Rejection rates ($\alpha = 0.05$) and nominal levels (only for global test)

Test	n	$C=3$ Glob	δ 1	2	3	$C=5$ Glob	δ 1	2	3	$C=7$ Glob	δ 1	2	3	H_0 (nominal level) Glob. test $C=3$ $C=5$ $C=7$
						Normal errors								
F	6	0.485	0.034	0.175	0.393	0.532	0.020	0.107	0.309	0.557	0.024	0.057	0.245	0.050 0.043 0.045
	10	0.813	0.092	0.377	0.761	0.823	0.036	0.236	0.632	0.832	0.048	0.127	0.533	0.043 0.054 0.050
	20	0.982	0.191	0.730	0.977	0.993	0.100	0.614	0.968	0.996	0.140	0.357	0.941	0.047 0.056 0.060
Friedman	6	0.382	0.017	0.094	0.292	0.406	0.005	0.043	0.164	0.440	0.005	0.020	0.127	0.044 0.041 0.030
	10	0.701	0.032	0.220	0.609	0.721	0.009	0.116	0.454	0.736	0.018	0.058	0.363	0.049 0.048 0.048
	20	0.959	0.066	0.530	0.935	0.975	0.043	0.417	0.878	0.987	0.078	0.223	0.878	0.045 0.047 0.054
Mean AR	6	0.415	0.058	0.206	0.365	0.442	0.025	0.107	0.277	0.475	0.022	0.055	0.206	0.051 0.052 0.038
	10	0.714	0.104	0.370	0.657	0.744	0.033	0.202	0.536	0.751	0.040	0.105	0.455	0.050 0.052 0.058
	20	0.964	0.174	0.651	0.943	0.975	0.086	0.532	0.912	0.987	0.127	0.315	0.899	0.048 0.050 0.057
Permutation	6	0.311	0.018	0.083	0.185	0.347	0.007	0.031	0.060	0.363	0.008	0.016	0.048	0.030 0.033 0.032
	10	0.738	0.091	0.350	0.643	0.721	0.027	0.135	0.357	0.729	0.036	0.073	0.301	0.044 0.039 0.049
	20	0.985	0.259	0.761	0.973	0.988	0.113	0.566	0.935	0.991	0.158	0.329	0.909	0.048 0.055 0.047
						Exponential errors								
F	6	0.567	0.057	0.252	0.489	0.571	0.024	0.123	0.375	0.584	0.030	0.069	0.310	0.040 0.045 0.046
	10	0.815	0.104	0.390	0.756	0.812	0.049	0.253	0.620	0.846	0.058	0.141	0.563	0.046 0.051 0.046
	20	0.980	0.211	0.728	0.972	0.991	0.103	0.600	0.955	0.991	0.144	0.369	0.931	0.050 0.044 0.043
Friedman	6	0.554	0.018	0.092	0.460	0.597	0.005	0.054	0.326	0.637	0.009	0.027	0.261	0.035 0.027 0.029
	10	0.850	0.041	0.241	0.785	0.904	0.014	0.188	0.680	0.936	0.036	0.111	0.618	0.054 0.056 0.039
	20	0.997	0.139	0.657	0.993	1.000	0.050	0.603	0.987	1.000	0.134	0.400	0.979	0.049 0.036 0.048
Mean AR	6	0.596	0.137	0.296	0.545	0.643	0.035	0.177	0.462	0.684	0.046	0.095	0.404	0.042 0.039 0.040
	10	0.861	0.207	0.470	0.820	0.917	0.059	0.360	0.769	0.943	0.087	0.221	0.734	0.059 0.067 0.045
	20	0.997	0.325	0.840	0.995	1.000	0.140	0.767	0.993	1.000	0.222	0.538	0.990	0.053 0.043 0.052
Permutation	6	0.421	0.028	0.155	0.260	0.449	0.012	0.033	0.085	0.478	0.016	0.018	0.074	0.026 0.031 0.027
	10	0.900	0.156	0.486	0.792	0.914	0.058	0.253	0.545	0.940	0.086	0.155	0.520	0.056 0.056 0.043
	20	0.988	0.314	0.768	0.966	0.994	0.150	0.602	0.917	0.995	0.238	0.422	0.914	0.053 0.042 0.050
						Student's t errors								
F	6	0.189	0.019	0.050	0.134	0.165	0.006	0.026	0.067	0.139	0.006	0.008	0.029	0.036 0.033 0.040
	10	0.261	0.037	0.079	0.203	0.263	0.011	0.042	0.133	0.222	0.006	0.015	0.072	0.030 0.026 0.039
	20	0.478	0.052	0.195	0.395	0.430	0.017	0.080	0.237	0.379	0.012	0.033	0.144	0.030 0.025 0.053
Friedman	6	0.216	0.013	0.044	0.154	0.197	0.004	0.017	0.071	0.234	0.003	0.010	0.052	0.046 0.034 0.046
	10	0.352	0.031	0.101	0.254	0.397	0.008	0.049	0.183	0.388	0.007	0.022	0.115	0.039 0.034 0.051
	20	0.693	0.058	0.265	0.624	0.746	0.024	0.155	0.501	0.790	0.030	0.086	0.443	0.050 0.042 0.056
Mean AR	6	0.241	0.049	0.091	0.192	0.232	0.011	0.045	0.116	0.273	0.013	0.022	0.086	0.051 0.043 0.054
	10	0.372	0.068	0.155	0.294	0.438	0.018	0.082	0.231	0.422	0.014	0.034	0.162	0.047 0.042 0.061
	20	0.703	0.095	0.316	0.641	0.756	0.040	0.209	0.543	0.802	0.047	0.111	0.495	0.057 0.046 0.061
Permutation	6	0.155	0.018	0.029	0.089	0.146	0.006	0.014	0.029	0.172	0.007	0.007	0.022	0.029 0.028 0.039
	10	0.403	0.066	0.155	0.312	0.428	0.014	0.057	0.171	0.412	0.011	0.024	0.120	0.039 0.033 0.040
	20	0.550	0.096	0.266	0.458	0.559	0.032	0.134	0.329	0.593	0.037	0.071	0.300	0.039 0.045 0.051

It is interesting to observe that not all procedures agree to reject the global null hypothesis ($\alpha = 0.05$). Moreover, the use of different combining functions for permutation tests seems to provide decision rules which are potentially more or less powerful.

It can be proved that the combined permutation test obtained using Fisher, Liptak or Tippet combining functions are so called 'admissible' combination, i.e. it does not exist any other type of combination which is uniformly more powerful. Note that if several combining functions are admissible they are equivalent as well.

Table 11.2 Category ratings for meat flavour for three breeds of geese

Consumer	X	Y	Z
1	3	2	3
2	4	5	4
3	3	2	3
4	1	4	2
5	2	4	2
6	1	3	3
7	2	5	4
8	2	5	2

Table 11.3 Category ratings for meat flavour for three breeds of geese

Test	Global	Pairwise comparisons		
		X vs. Y	X vs. Z	Y vs. Z
F	0.028	0.026	0.675	0.292
Friedman	0.152	–	–	–
Mean AR Permutation	0.158	–	–	–
Fisher	0.049	0.048	0.235	0.113
Tippet	0.107	–	–	–
Liptak	0.019	0.026	0.256	0.103

11.6 Conclusions

In this chapter we have presented a combination-based permutation solution for hypothesis testing within the framework of randomized complete block design. The proposed solution may suggest to practitioners in the field of evaluation for educational services and quality of products an effective approach, especially when using ordered categorical variables, such as in the case of sensorial evaluations. As confirmed by the presented simulation study, the nonparametric tests are certainly good alternatives, in particular respect to the traditional parametric F and t test. In fact, even in case of normality, the power of permutation tests is nearly the same as that of the parametric tests, while in case of asymmetric or heavy tailed error distributions permutation tests can provide higher power. Hence, in each practical situation where the normality assumption is hard to justify, the proposed nonparametric procedure can be considered a valid solution.

Finally, as suggested by the real case study, a possible way to improve power of permutation tests is to better investigate the role of the combining functions. Note that our proposed permutation test applies a combining function two times: at first in order to combine the partial pairwise permutation tests to obtain a global test, then we apply a combining function in order to perform a suitable multiplicity correction strategy for pairwise permutation p-values.

Chapter 12
A permutation test for umbrella alternatives

Dario Basso, Fortunato Pesarin and Luigi Salmaso

12.1 Introduction

There is a wide variety of stochastic ordering problems where J groups (typically ordered with respect to time) are observed along with a (continuous) response. The interest of the study may be on finding the change-point group, i.e. the group where an inversion of trend of the variable under study is observed. A change point is not merely a maximum (or a minimum) of the time-series function, but a further requirement is that the trend of the time-series is monotonically increasing before that point, and monotonically decreasing afterwards. A suitable solution can be provided within a conditional approach, i.e. by considering some suitable nonparametric combination of dependent tests for simple stochastic ordering problems.

In a one-way ANOVA experiments, it is common that the response variable increases with an increase in the treatment level up to a point, then decreases with further increase in the treatment level. In the literature, this up-then-down pattern has been identified as umbrella ordering (Mack and Wolfe, 1981). Umbrella orderings can be observed with many physical and biological phenomena in a wide variety of scientific research areas.

There has been considerable previous work on procedures designed to test homogeneity against umbrella ordering alternatives. Such testing procedures are generally based on ranks. An introductory review on such tests can be found in Wolfe (2006)

Dario Basso
Department of Management and Engineering, University of Padua, Stradella S. Nicola 4, 36100 Vicenza, Italy,
e-mail: basso@gest.unipd.it

Fortunato Pesarin
Department of Statistics, University of Padua Via Cesare Battisti 241, 35121 Padua, Italy,
e-mail: pesarin@stat.unipd.it

Luigi Salmaso
Name, Department of Management and Engineering, University of Padua, Stradella S. Nicola 4, 36100 Vicenza, Italy, e-mail: salmaso@gest.unipd.it

and in Millen and Wolfe (2005). Mack and Wolfe (1981) proposed a test statistic based on the Jonckheere-Terpstra statistic.

Hettmansperger and Norton (1987) pointed out that no comparisons are made in the Mack-Wolfe test between samples preceding the known peak and those following it. Pan (1996) proposed to retrieve the information across the peak using a test statistic which is the maximum of the Jonckheere-Terpstra statistics. However, within this kind of alternatives, tests for umbrella alternatives with an unknown peak are much more practical. The most common approach to construct test statistics for this setting is to take the maximum of the test statistics for umbrella alternatives with known peaks (Mack and Wolfe, 1981; Hettmansperger and Norton, 1987; Chen and Wolfe, 1990; Shi, 1998; Hartlaub and Wolfe, 1999). Magel and Qin (2003) proposed a test that extends the Chen and Wolfe (1990) test for umbrella alternatives with an unknown peak to use with ranked-set samples data which is essentially based on the procedure proposed by Hartlaub and Wolfe (1999). The proposed test, however, is not the best in situations where the first location shift or the last location shift is much higher than the others. If the first or last location shift is expected to be much higher than the remaining shifts, and the remaining location shifts are expected to be approximately equal, both the Chen–Wolfe and Mack–Wolfe tests are recommended for use. It has also been shown (Kössler, 2006) that generally the Hettmansperger-Norton-type test performs better, densely followed by the Chen–Wolfe-type test and the Shi-type test. Recently, Pan (2008) proposed a non-parametric distribution-free confidence procedure for umbrella orderings by constructing a random confidence subset of the ordered treatments such that it contains all the unknown peaks (optimal treatments) of an umbrella ordering with any pre-specified confidence level. Anyway in the literature it is well recognized that Mack and Wolfe and Chen and Wolfe type tests are the milestones for the umbrella alternative problems.

There are very few papers concerning nonparametric permutation proposals for umbrella alternatives a part for some hints given in Manly (1997) and the recent paper by Neuhäuser et al. (2003) where a modified Jonckheere–Terpstra test is presented in a suitable permutation version in order to obtain reliable results with small, sparse, unbalanced, and tied data.

In this chapter, we introduce a permutation test for umbrella alternatives. Permutation tests do not require assumption on the distribution of data. Moreover, the distribution of the test statistic is exact, whereas the majority of existing tests for umbrella alternatives are exact only asymptotically. Therefore, permutation tests can be applied at any α-values, whereas the existing competitors require tabulated critical values for some α-levels that have been chosen by the related authors. The procedure we are introducing works even with very small sample sizes (say 2 replicates for each treatment) and/or in unbalanced cases. Thus, we recommend this procedure when small sample sizes are available or when data cannot be assumed to follow a specific distribution.

The procedure proposed makes use of the nonparametric combination, introduced by Pesarin (2001). This methodology is based on the decomposition of complex hypotheses (such as the umbrella alternative) into a set of simple "partial" hypotheses. Each partial hypothesis is then tested by a suitable "partial test" and the

12 Permutation test for umbrella alternatives

information related to partial tests is then combined together through the nonparametric combination leading to a global test statistic for the complex problem.

The context is that of one-way ANOVA experiments, where the experimental factor (time, increasing doses of drug) levels determines the treatments which identify the J groups.

Let Y_{ij} be the observed response variable on the ith subject from group $j = 1, \ldots, J$. We assume Y_{ij} to follow the additive model:

$$Y_{ij} = \mu + \delta_j + \varepsilon_{ij} \qquad i = 1, \ldots, n_j, \qquad (12.1)$$

where μ is the population mean, δ_j is the treatment effect on the jth group (which may also be stochastic) and ε_{ij} are exchangeable errors with zero mean and finite variance σ^2 (typically i.i.d. random variables, independent of δ_j's), and n_j are fixed sample sizes. Let $F_j(y)$ be the cumulative distribution function of the response variable in group j. Then we wish to assess the null hypothesis of no treatment effect:

$$H_0: F_1(y) = F_2(y) = \cdots = F_J(y) \qquad \forall y \in \mathbb{R},$$

against the umbrella alternative hypothesis:

$$H_1: F_1(y) \geq \cdots \geq F_{j-1}(y) \geq F_j(y)$$
$$\leq F_{j+1}(y) \leq \cdots \leq F_J(y)$$

for some $j \in \{1, \ldots, J\}$, and with at least one strict inequality. That is, the interest of the study is on finding the change-point group j (if it exists), i.e. the group where an inversion of trend of the variable under study is observed. A change point is not merely a maximum of the time-series function. Actually, a further requirement is that the trend of the time-series is monotonically alternative before group j and monotonically non increasing afterwards. Thus there are two main aspects to consider: (i) is there any umbrella behaviour due to the experimental factor? (ii) If so, which one is the change-point group? A parametric solution to this problem is very difficult, especially when $J > 2$. These hypotheses define a problem of *isotonic inference* (see Hirotsu, 1998).

The proposal of this work is a conditional approach to the observed data. If we knew the peak group, say the $\hat{j}$th one, the problem of umbrella alternatives could be simplified in an intersection of alternative hypotheses:

$$H_1 = H_{1\hat{j}}^{\nearrow} \cap H_{1\hat{j}}^{\searrow}$$

where:

$$H_{1\hat{j}}^{\nearrow} = F_1(y) \cdots \geq F_{\hat{j}-1}(y) \geq F_{\hat{j}}(y)$$
$$H_{1\hat{j}}^{\searrow} = F_{\hat{j}}(y) \leq F_{\hat{j}+1}(y) \leq \cdots \leq F_J(y).$$

That is, if the peak group were known, the umbrella alternative could be written as the intersection of two simple stochastic ordering alternatives (an increasing one and a decreasing one). In order to introduce the permutation test for umbrella alternative, we first need to introduce some suitable permutation tests for ordinary stochastic ordering problems (Sect. 12.2). This will latter require the nonparametric combination (NPC) methodology, which is a useful tool when one needs to combine different informations/aspects of the same problem. The NPC methodology is introduced in Chap. 1 and it will be applied for defining the permutation test for simple stochastic ordering problems. Section 12.4 of this chapter is dedicated to our proposal for umbrella problems. In Sect. 12.4 the test proposed is evaluated through a simulation study, and compared with that of Mack & Wolfe. In Sect. 12.5 an application example is discussed.

12.2 Simple stochastic ordering alternatives

Under the assumption of model (12.1), let us consider the simple stochastic ordering problem for the first $\hat{j}$ samples to assess the null hypothesis $F_1(y) = F_2(y) = \cdots = F_{\hat{j}}(y)$ against the alternative hypothesis $F_1(y) \geq \cdots \geq F_{\hat{j}-1}(y) \geq F_{\hat{j}}(y)$. Note that under the null hypothesis the elements of the response are exchangeable (this fact enables us to provide the null distribution of a proper test statistic).

If $\hat{j} = 2$, the stochastic ordering problem reduces to a two-sample problem with restricted alternative. If $\hat{j} > 2$, then let us consider the whole data set is split into two pooled pseudo-groups, where the first is obtained by pooling together data of the first k groups (ordered with respect to the treatment levels), and the second by pooling together the remaining observations. In order to better understand the reason why we pool together the ordered groups, suppose $\hat{j} = 3$ and let us consider the following theorem:

Theorem 1: Let X_1, X_2, X_3 be mutually independent random variables which admit cumulative distribution function $F_j(t)$, $t \in \mathbb{R}$, $j = 1,2,3$. Then, if $X_1 \stackrel{d}{\leq} X_2 \stackrel{d}{\leq} X_3$, we have:

(i) $X_1 \stackrel{d}{\leq} X_2 \oplus X_3$ and (ii) $X_1 \oplus X_2 \stackrel{d}{\leq} X_3$,

where $W \oplus V$ indicates a mixture of random variables W and V, i.e. $F_{W \oplus V}(t) = \omega_W F_W(t) + \omega_V F_V(t)$, $t \in \mathbb{R}$, $\omega_W, \omega_V \in [0,1]$, $\omega_W + \omega_V = 1$.

Proof: By definition, $X_1 \stackrel{d}{\leq} X_2 \stackrel{d}{\leq} X_3$ is equivalent to $F_1(t) \geq F_2(t) \geq F_3(t)$, $\forall t \in \mathbb{R}$. The random variable $X_1 \oplus X_2$ has cumulative distribution function equal to:

$$F_{X_1 \oplus X_2}(t) = \omega_1 F_1(t) + \omega_2 F_2(t),$$

with $\omega_1, \omega_2 \in [0,1]$, $\omega_1 + \omega_2 = 1$. Therefore, by hypothesis:

$$F_{X_1 \oplus X_2}(t) = \omega_1 F_1(t) + \omega_2 F_2(t)$$
$$\geq \omega_1 F_2(t) + \omega_2 F_2(t) = F_2(t),$$

so $X_1 \oplus X_2 \overset{d}{\leq} X_2$ and we have proved (ii). In the same way, let $F_{X_2 \oplus X_3}(t) = \omega_2 F_{X_2}(t) + \omega_3 F_{X_3}(t)$ with $\omega_2, \omega_3 \in [0,1]$, $\omega_2 + \omega_3 = 1$, then:

$$F_{X_2 \oplus X_3}(t) = \omega_2 F_2(t) + \omega_3 F_3(t)$$
$$\leq \omega_2 F_2(t) + \omega_3 F_2(t) = F_2(t),$$

therefore $X_2 \oplus X_3 \overset{d}{\geq} X_2$, and this proves (i).

Now, conditionally to the observed data, consider the pooled vector of observations $\mathbf{y}_1 \uplus \mathbf{y}_2 = [\mathbf{y}_1, \mathbf{y}_2]'$, where $\mathbf{y}_j$ is a vector of n_j observations from $F_{Y_j}(y)$, $j = 1, 2$, and the symbol $\uplus$ denotes the pooling of two vectors. Then the random variable $Y_1 \oplus Y_2$ describing the generic observation of $\mathbf{y}_1 \uplus \mathbf{y}_2$ has (empirical) cumulative distribution function equal to:

$$\hat{F}_{Y_1 \oplus Y_2}(y) = \frac{1}{n_1 + n_2} \sum_{j=1}^{2} \sum_{i=1}^{n_j} I(y_{ij} \leq y)$$
$$= \frac{n_1}{n_1 + n_2} \frac{\sum_{i=1}^{n_1} I(y_{i1} \leq y)}{n_1} + \frac{n_2}{n_1 + n_2} \frac{\sum_{i=1}^{n_2} I(y_{i2} \leq y)}{n_2}$$
$$= \omega_1 \hat{F}_{Y_1}(y) + \omega_2 \hat{F}_{Y_2}(y),$$

where $I(\cdot)$ is the indicator function. Therefore, conditionally, $Y_1 \oplus Y_2$ has a mixture distribution.

By extending this result to the $\hat{j}$ groups and by applying Theorem 1, we have that if $Y_1 \overset{d}{\leq} Y_2 \overset{d}{\leq} \cdots \overset{d}{\leq} Y_{\hat{j}}(y)$ holds, then:

$$Y_{1 \oplus 2 \oplus \cdots \oplus k} \overset{d}{\leq} Y_{k+1 \oplus k+2 \oplus \cdots \oplus \hat{j}} \quad \forall k \in \{1, \ldots, \hat{j}-1\}.$$

In general, let $\mathbf{z}_{1(k)} = \mathbf{y}_1 \uplus \mathbf{y}_2 \uplus \cdots \uplus \mathbf{y}_k$ be the first and $\mathbf{z}_{2(k)} = \mathbf{y}_{k+1} \uplus \cdots \uplus \mathbf{y}_{\hat{j}}$ be the second (ordered) pseudo-group, $k = 1, \ldots, \hat{j} - 1$. Let $Z_{1(k)}$ and $Z_{2(k)}$ be the random variables describing the generic observation of the pooled vectors $\mathbf{z}_{1(k)}$ and $\mathbf{z}_{2(k)}$, respectively. In the null hypothesis, data of every pair of pseudo-groups are exchangeable because the related variables satisfy the relationships $Z_{1(k)} \overset{d}{=} Z_{2(k)}$, $k = 1, \ldots, \hat{j} - 1$. In the alternative, by Theorem 1, we have $Z_{1(k)} \overset{d}{\leq} Z_{2(k)}$, which corresponds to the monotonic stochastic ordering (dominance) between any pair of pseudo-groups (i.e. for $k = 1, \ldots, \hat{j} - 1$). This suggests that we express the hypotheses in the equivalent form:

$$H_0 : \left\{ \bigcap_{k=1}^{\hat{j}-1} (Z_{1(k)} \stackrel{d}{=} Z_{2(k)}) \right\},$$

against

$$H_{1\hat{j}}^{\nearrow} : \left\{ \bigcup_{k=1}^{\hat{j}-1} (Z_{1(k)} \stackrel{d}{\leq} Z_{2(k)}) \right\},$$

where a breakdown into a set of sub-hypotheses (or partial hypotheses) is emphasized.

Let us pay attention to the kth sub-hypothesis $H_{0k} : \{Z_{1(k)} \stackrel{d}{=} Z_{2(k)}\}$ against $H_{1k} : \{Z_{1(k)} \stackrel{d}{\leq} Z_{2(k)}\}$. Note that the related sub-problem corresponds to a two-sample comparison for restricted alternatives, a problem which has an exact and unbiased permutation solution (for further details see Pesarin, 2001). This solution is based on the test statistics (among others):

$$T_{k\nearrow} = \frac{\bar{Z}_{2(k)} - \bar{Z}_{1(k)}}{\sqrt{\hat{\sigma}_k^2 \left(\frac{1}{n_{1(k)}} + \frac{1}{n_{2(k)}} \right)}} \qquad k = 1, \ldots, \hat{j}-1, \qquad (12.2)$$

where $\bar{Z}_{2(k)}$ and $\bar{Z}_{1(k)}$ are sample means of the second and the first pseudo-group, respectively, $\hat{\sigma}_k^2$ is the pooled estimate of the error variance, and $n_{1(k)}$ and $n_{2(k)}$ are the lengths of $\mathbf{z}_{1(k)}$ and $\mathbf{z}_{2(k)}$. Large values of the test statistics $T_{k\nearrow}$ are significant against $H_{0k} : Z_{1(k)} \stackrel{d}{=} Z_{2(k)}$ in favor of the alternatives $H_{1k}^{\nearrow} : Z_{1(k)} \stackrel{d}{\leq} Z_{2(k)}$. We can obtain a permutation test for each sub-problem H_{0k} vs. $H_{1k}^{\nearrow}$ by the following algorithm:

- Let $\mathbf{y} = [\mathbf{y}_1, \mathbf{y}_2, \ldots, \mathbf{y}_{\hat{j}}]'$ be the vector of the observed data in $\hat{j}$ groups.
- for $k = 1, \ldots, \hat{j}-1$, repeat:
  1. Let $\mathbf{z}_{1(k)} = [\mathbf{y}_1, \ldots, \mathbf{y}_k]'$ and $\mathbf{z}_{2(k)} = [\mathbf{y}_{k+1}, \ldots, \mathbf{y}_{\hat{j}}]'$;
  2. Compute the observed values of the partial test statistics for the sub-problem H_{0k} vs. $H_{1k}^{\nearrow}$ by computing:

$$T_{k\nearrow} = \frac{\bar{z}_{2(k)} - \bar{z}_{1(k)}}{\sqrt{\hat{\sigma}_k^2 \left(\frac{1}{n_{1(k)}} + \frac{1}{n_{2(k)}} \right)}}. \qquad (12.3)$$

- Consider a large number B of independent random permutations of the response $\mathbf{y}$, and let $\mathbf{y}^*_b$ be a random permutation of $\mathbf{y}$. At each step $b = 1, \ldots, B$, repeat:
  1. let $\mathbf{z}^*_{1(k)}$ be the vector with the first $n_{1(k)} = \sum_{\ell=1}^{k} n_\ell$ observations and $\mathbf{z}^*_{2(k)}$ be the vector of the last $n_{2(k)} = \sum_{\ell=k+1}^{\hat{j}} n_\ell$ observations of $\mathbf{y}^*_b$;
  2. Obtain the permutation null distribution of the test statistic by computing:

$$ {}^b T^*_{k\nearrow} = \frac{\bar{z}^*_{2(k)} - \bar{z}^*_{1(k)}}{\sqrt{\hat{\sigma}^{*2}_k \left(\frac{1}{n_{1(k)}} + \frac{1}{n_{2(k)}} \right)}}, $$

where $\bar{z}^*_{2(k)}$ and $\bar{z}^*_{1(k)}$ are the means of $\mathbf{z}^*_{2(k)}$ and $\mathbf{z}^*_{1(k)}$, respectively, and $\hat{\sigma}^{*2}_k$ is the pooled estimate of the error variance. Thus, the set $\left\{ {}^b T^*_{k\nearrow}, b = 1, \ldots, B \right\}$ is a random sample from the null permutation distribution of the test statistic $T_{k\nearrow}$.

- Obtain the *p*-value of each sub-problem (partial *p*-value) by computing:

$$ p_{k\nearrow} = \frac{\# \left[T^*_{k\nearrow} \geq T_{k\nearrow} \right]}{B}. $$

The previous algorithm provides $\hat{j} - 1$ *p*-values related to the sub-hypothesis system H_{0k} against $H^{\nearrow}_{1k}$. In order to combine the partial information into a global test we require the NPC methodology which is introduced in Chap. 1. Obviously, if the alternative hypothesis is:

$$ H^{\searrow}_{1j} : F_{\hat{j}}(y) \leq F_{\hat{j}+1}(y) \cdots \leq F_J(y), $$

the previous algorithm still apply by replacing the test statistic (12.2) with:

$$ T_{k\searrow} = \frac{\bar{Z}_{1(k)} - \bar{Z}_{2(k)}}{\sqrt{\hat{\sigma}^2_k \left(\frac{1}{n_{1(k)}} + \frac{1}{n_{2(k)}} \right)}}, \quad k = \hat{j}, \ldots, J-1. $$

12.3 Permutation test for umbrella alternatives

If the peak group were known, then the umbrella alternative could be detected by combining together two partial tests for simple stochastic ordering alternatives. However, it is generally unknown. Nevertheless we can detect the peak group by repeating the procedure for known peak *as if* every group were the known peak group: that is, for each $j \in 1, \ldots, J$. Let:

$$ \psi_{j\nearrow} = \sum_{k=1}^{j} T_{k\nearrow} \quad \text{and} \quad \psi_{j\searrow} = \sum_{k=j}^{J-1} T_{k\searrow} $$

be two partial test to assess $H_{0j} : F_1(y) = F_2(y) = \cdots = F_K(y)$ against respectively $H^{\nearrow}_{1j}$ and $H^{\searrow}_{1j}$ by applying the *direct* nonparametric combination of the partial tests $T_{k\nearrow}$'s and $T_{k\searrow}$'s, then:

- Obtain the partial p-values to assess H_{0j} against $H_{1j}^{\nearrow}$ and $H_{1j}^{\searrow}$, respectively. Let $\left(p_{j\nearrow}^{G}, p_{j\searrow}^{G}\right)$ be the pair of p-values from the observed data.
- Obtain null distribution of the pair of p-values to assess H_{0j} against respectively $H_{1j}^{\nearrow}$ and $H_{1j}^{\searrow}$. This will be indicated with the pair $\left({}^{b}p_{j\nearrow}^{G*}, {}^{b}p_{j\searrow}^{G*}\right)$, $b = 1, \ldots, B$. That is, $\left({}^{b}p_{j\nearrow}^{G*}, {}^{b}p_{j\searrow}^{G*}\right)$ is obtained by applying the previous algorithm for simple stochastic ordering alternatives and by replacing $\mathbf{y}$ with $\mathbf{y}_b^*$;
- Obtain the observed value of the test statistic with *Fisher's* NPC function:

$$\Psi_j = -2\log\left(p_{j\nearrow}^{G} \cdot p_{j\searrow}^{G}\right).$$

- Obtain the null distribution of Ψ_j by computing:

$${}^{b}\Psi_j^* = -2\log\left({}^{b}p_{j\nearrow}^{G*} \cdot {}^{b}p_{j\searrow}^{*G}\right), \qquad b = 1, \ldots, B.$$

- Obtain the p-value for umbrella alternative on group k as:

$$\pi_j = \frac{\#\left[{}^{b}\Psi_j^* \geq \Psi_j\right]}{B}.$$

Note that if π_j is significant then there is evidence on data of an umbrella alternative with peak group j. In order to evaluate if there is a significant presence of any umbrella alternative, we finally combine the p-values for umbrella alternative of each group.
- To do so: obtain the null distribution of the p-value for umbrella alternative on group j as:

$${}^{b}\pi_j^* = \frac{\#\left[\Psi^* \geq {}^{b}\Psi_j^*\right]}{B}, \qquad b = 1, \ldots, B,$$

where Ψ^* is the vector with the permutation null distribution of Ψ_j.
- Apply *Tippett's* combining function to the π_j's, providing the observed value of the global test statistic for umbrella alternative in any group as:

$$\Pi = \min(\pi_1, \pi_2, \ldots, \pi_J).$$

Note that small values of Π are significant against the null hypothesis.
- Obtain the null distribution of Π^G by computing:

$${}^{b}\Pi^* = \min\left({}^{b}\pi_1^*, {}^{b}\pi_2^*, \ldots, {}^{b}\pi_K^*\right), \qquad b = 1, \ldots, B.$$

- Obtain the global p-value as:

$$\Pi^G = \frac{\#[{}^{b}\Pi^* \leq \Pi]}{B}.$$

Note that the combining functions are applied simultaneously to each random permutation, providing the null distributions of partial and global tests as well. The NPC methodology applies three times in this testing procedure:

1. When obtaining simple stochastic ordering tests to assess $H_{1k}^{\nearrow}$ and $H_{1k}^{\searrow}$ for the kth group ("direct" combining function).
2. when combining together two partial tests for simple stochastic ordering alternatives, providing a test for umbrella for each group as it were the known peak group ("Fisher's" combining function).
3. when combining together the partial tests for umbrella on each group ("Tippett's" combining function).

A significant global p-value Π^G indicates that there is evidence in favor of an umbrella alternative. The peak group is then identified by looking at the partial p-values for umbrella alternatives $\{\pi_1, \pi_2, \ldots, \pi_k, \ldots, \pi_K\}$. The peak group (if any) is then identified as the one with minimum p-value.

The proposed algorithm may still apply with different combining functions in first two steps, but not in the third step. This is because Tippett's ψ is significant only when at least one of its arguments is significant. As regards our choices in the first two steps of the algorithm, the direct combining function in step 1 has been chosen for computational reasons, whereas Fisher's combining function in step 2 has been applied because Fisher's ψ is generally suitable when no specific knowledge of the sub-alternatives is expected.

12.4 Simulation study

In this section we show the performances of the permutation test for umbrella alternatives by providing results on some simulations under the null hypothesis and under some alternatives. The chosen settings are $K = 5$ groups with $n_j = 3$ observations each ($j = 1, \ldots, 5$). The simulated data have a standard normal distribution, possibly with some non random location shifts in some groups (under the alternative hypothesis). The location shifts considered in each simulation are indicated by the symbols δ_k on the top of each table. Each simulation is based on 1,000 independent Monte Carlo data generations. The simulation settings have been chosen in accordance with the example appeared in Mack and Wolfe (1981). This example is about a score intelligence test: five male groups with three subjects each were evaluated through the Welchsler Adult Intelligence Scale (WAIS). The groups were identified by different classes of age. The authors conjectured that the intelligence score follows an umbrella trend. Figure 12.1 shows the boxplot representation of data, and the dotted line is the trend line connecting the group means. For details on the test statistic and data of the example we refer to Mack and Wolfe (1981). The data of Example 1 in Mack and Wolfe were analyzed by both the permutation test and Mack and Wolfe's test. Mack and Wolfe's test gave pretty significant results in favor of the umbrella alternative, providing an approximated p-value equal to 0.0328

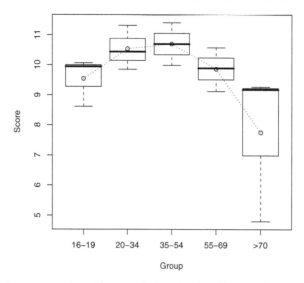

Fig. 12.1 Boxplot representation of the example from Mack and Wolfe (1981).

with the third age-group as the estimated peak group. The results of the permutation test are shown in Table 12.1: the global p-value is equal to 0.014, indicating a strongly significant presence of an umbrella alternative. The peak group is then individuated through the partial p-values: provided that the global test is significant, the peak group is the one with minimum partial p-value. In this example the third group ($\pi_3 = 0.00299$) is the peak group.

Table 12.1 Permutation Test results, WAIS score data

Age	15–19	20–34	35–54	55–69	> 70
π_k	0.05794	0.00599	**0.00299**	0.12687	0.94306
		$\Pi^G = \mathbf{0.014}$			

Table 12.2 reports the results of a simulation under H_0. The rejection rates of the null hypothesis of partial and global tests at different α-sizes are shown. Note how the rejection rates of the global test column (indicated by "G.T.") are close to the nominal ones. Then, for each group, the rejection rates of the partial tests for peak-known umbrella are also shown. In order to account for multiplicity, the partial p-values are compared to the adjusted α-level through a Bonferroni's correction (therefore the actual nominal level for the partial tests is $\alpha/5$). Under the null hypothesis, the probability of observing a peak group should be uniformly distributed among the K groups, therefore the estimated probabilities of the event

12 Permutation test for umbrella alternatives

$\pi_k = \min\{p_1, \ldots, p_K\}$ conditional to the rejection of the global null hypothesis are also shown.

Table 12.2 Rejection rates of the permutation test under H_0

Group	1	2	3	4	5	
δ_k	0	0	0	0	0	
α			Partial tests			G.T.
0.05	0.006	0.014	0.010	0.012	0.012	0.046
0.10	0.028	0.024	0.020	0.028	0.030	0.100
0.20	0.042	0.050	0.044	0.056	0.046	0.192
α			$\Pr\{\pi_k = \min_j \pi_j \| \Pi^G \leq \alpha\}$			Tot.
0.05	0.206	0.176	0.206	0.235	0.176	1
0.10	0.196	0.214	0.196	0.196	0.196	1
0.20	0.208	0.188	0.208	0.198	0.198	1

Table 12.3 shows the behaviour of the test under an umbrella alternative with peak on third group. Note that the rejection rates of the global test are far bigger than the nominal levels. Moreover, the rejection rates of the partial tests (accounting for multiplicity) are directly proportional to the sizes of δ_k's, and that group 3 has been detected as the peak group about 35% of the times that the global null hypothesis has been rejected at all α levels (however note that the nonzero δ_k sizes are very close to each other and to the variance of data distribution $\sigma^2 = 1$).

Table 12.3 Rejection rates of the permutation test under umbrella alternative

Group	1	2	3	4	5	
δ_k	0	0.9	1	0.9	0	
α			Partial tests			G.T.
0.05	0.002	0.116	0.128	0.094	0.004	0.306
0.10	0.006	0.166	0.216	0.144	0.008	0.446
0.20	0.014	0.264	0.290	0.226	0.030	0.622
α			$P\{\pi_k = \min_j \pi_j \| \Pi^G \leq \alpha\}$			Tot.
0.05	0.007	0.366	0.366	0.248	0.013	1
0.10	0.018	0.332	0.386	0.247	0.018	1
0.20	0.023	0.354	0.354	0.241	0.029	1

In Table 12.4 we have set the location shifts in order to simulate an anti-umbrella alternative. That is, data are not under the null hypothesis, but the true alternative hypothesis is not of umbrella kind. The trend is first decreasing then increasing, and the permutation test should not recognize this kind of alternative, since it has been specifically created for umbrella alternatives. Indeed, the rejection rates of the global tests are always lower than the related nominal levels.

Table 12.4 Rejection rates of the permutation test under anti-umbrella alternative

Group	1	2	3	4	5	
δ_k	1	0.5	0	0.5	1	
α	Partial tests					G.T.
0.05	0.010	0.000	0.000	0.000	0.015	0.022
0.10	0.020	0.002	0.000	0.000	0.018	0.040
0.20	0.027	0.004	0.000	0.000	0.028	0.068
α	$P\{\pi_k = \min_j \pi_j | \Pi^G \leq \alpha\}$					
0.05	0.455	0.000	0.000	0.000	0.545	1
0.1	0.500	0.050	0.000	0.000	0.450	1
0.2	0.441	0.088	0.000	0.000	0.471	1

12.5 Case study: graduates in engineering

We have applied the testing procedure described above to a dataset of session degrees at the School of Engineering of the University of Padova in the period 2003–2006. Data are 1529° from 16 Engineering Courses in period 2003–2006. The groups are defined by the session of degree (E1, E2, E5: summer sessions; A1: autumn sessios; I1, I2, I5, I6, I7, I8: winter sessions). Each Course has a different number of degree sessions (from 4 to 8), and students within the same degree session (even from different years) are treated as replicates.

The aim of the study is to find out if there are some courses where the degree evaluation has a significant umbrella trend and, if so, to determine the peak session. Our feeling is that the best students should take a bit to complete their degree thesis and therefore we expect to find them not in the early session (e.g. summer sessions E1, E2, E5 or autumn sessions A1).

We have run the permutation test for umbrella alternatives on the degree data for each Engineering Course and found the results shown in Table 12.5. There is a significant presence of the umbrella alternative in 7 courses over 16 at a nominal significance level $\alpha = 0.05$; All those courses which showed an umbrella trend on the degree evaluation did also indicate that the peak group was in session I1 or I2, that correspond to the first winter sessions. Note that the testing procedure may produce *ex aequo*, as in the Mechanical Engineering course (the peaks are I1 and I2). We have only emphasized (in bold face) those courses whose global p-value is less than 5%, although the early winter sessions have often provided singificant partial *p*-values, thus indicating that our feeling about when the best students take their degree was right. We have considered 1,000 permutation per each analysis.

Finally, let us explore in detail the degree evaluation trend of some courses in Table 12.5: Figs. 12.2, 12.3, 12.4, 12.5, 12.6, 12.7, 12.8 and 12.9 represent, for each course, a boxplot representation, and the dotted lines link the group means. In the title of each boxplot there is the name of the course and its global *p*-value. The partial *p*-values are displayed at the bottom, together with the group labels.

12 Permutation test for umbrella alternatives

Table 12.5 Results of the analysis on Engineering Courses. Global and partial p-values

Course	G.T.	E1	E2	E5	A1	I1	I2	I5	I6	I7	I8
Aereospatial	**0.006**	–	0.894	–	0.317	0.004	**0.001**	0.016	0.107	–	–
Automation	0.685	0.461	0.520	–	0.536	0.210	0.730	0.929	0.576	–	–
Biomedic	0.054	0.950	0.740	–	0.927	0.300	0.004	0.150	0.054	–	–
Building	**0.008**	0.994	0.738	1	0.526	**0.001**	0.018	0.010	0.007	–	–
Chemical	0.192	–	0.716	–	0.657	0.040	0.049	1	0.266	–	–
Civil	**0.008**	0.416	0.726	1	0.591	**0.001**	0.134	0.311	0.585	–	–
Computer Science	**0.046**	0.132	0.894	–	0.748	**0.007**	0.008	0.512	0.869	–	–
Electronic	**0.019**	0.541	0.124	–	0.263	0.005	**0.003**	0.369	0.459	–	–
Electrotechnical	0.344	1	0.945	–	0.917	0.170	0.410	0.942	1	–	–
Energetic	0.983	1	0.997	–	0.999	0.919	0.860	1	–	–	–
Environmental	0.153	0.486	0.362	0.520	0.831	0.025	0.112	0.530	0.556	–	–
Information	0.267	1	0.992	–	0.992	0.149	1	–	–	–	–
Management	**0.008**	0.949	0.808	0.004	0.219	**0.001**	0.003	0.021	0.059	–	–
Materials	0.389	–	0.124	–	0.560	1.000	0.934	–	–	–	–
Mechanical	**0.009**	0.405	0.628	0.106	0.098	**0.001**	**0.001**	0.059	0.542	0.169	0.617
Telecommunications	0.750	1	0.724	0.596	0.888	0.876	0.437	0.931	1	–	–

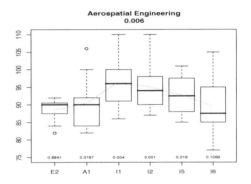

Fig. 12.2 Boxplot and results of the analysis of Aerospatial Engineering Course. The global p-values is shown on the *top* of the figure, the partial p-values are displayed at the *bottom*, in correspondence of each session's name.

In Aerospatial Engineering course (Fig. 12.2), there is a well-defined umbrella trend with peak group in early winter session I2. The global p-value is equal to 0.006, indicating evidence of an umbrella trend. In order to determine the peak group, we must look at the partial p-values: in this case the minimum p-value is that of session I2 ($p = 0.001$).

The degree evaluation trend is not well defined in Building Engineering course (Fig. 12.3). However, the global p-value is highly significant ($\Pi^G = 0.008$), and, visually, there seems to be two candidates as peak groups (I1 and I5). Note how group I1 has higher variance than group I5, although their sample means are very close to each other. The partial p-value of group I1 is equal to 0.001, that is more significant than that of group I5, therefore the peak group is situated in the first winter session.

Fig. 12.3 Boxplot and results of the analysis of Building Engineering Course. The global *p*-values is shown on the *top* of the figure, the partial *p*-values are displayed at the *bottom*, in correspondence of each session's name.

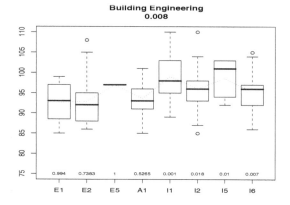

Fig. 12.4 Boxplot and results of the analysis of Chemical Engineering Course. The global *p*-values is shown on the *top* of the figure, the partial *p*-values are displayed at the *bottom*, in correspondence of each session's name.

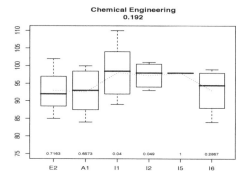

In the Chemical Engineering course (Fig. 12.4), instead, there seems to be no evidence of an umbrella trend, as there is a sort of plateau in winter sessions. The global *p*-value is not significant ($\Pi^G = 0.192$). Note, however, that the degree evaluations of early winter sessions are pretty higher than those of summer, autum, and late winter sessions. Thus, even if there is no significant evidence of a real umbrella trend, our feeling about student behaviour seems to be still valid. Note that in session I5 there was a single student, and the related *p*-value is equal to 1.

The global *p*-value of Civil Engineering course is highly significant ($\Pi^G = 0.008$). The boxplot representation (Fig. 12.5) is somehow misleading, since in session E5 there was only a single student. If we do not consider session E5 and look at the medians of each boxplot, there is graphical evidence of an umbrella trend with peak group in session I1 (its partial *p*-value is equal to 0.001).

In Computer Science Engineering course (Fig. 12.6), there is a pretty clear trend of degree evaluation if we do not consider group E1 (that is made of 5 students only). The global *p*-value is $\Pi^G = 0.046$, and the peak group is the first winter session I1, whose partial *p*-value is equal to 0.007.

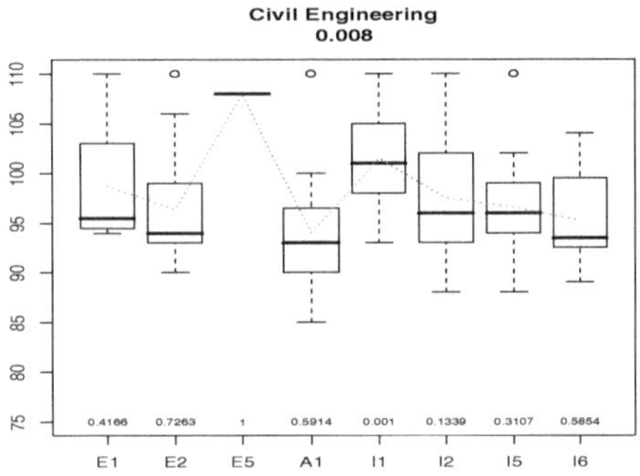

Fig. 12.5 Boxplot and results of the analysis of Civil Engineering Course. The global *p*-values is shown on the *top* of the figure, the partial *p*-values are displayed at the *bottom*, in correspondence of each session's name.

Fig. 12.6 Boxplot and results of the analysis of Computer Science Engineering Course. The global *p*-values is shown on the *top* of the figure, the partial *p*-values are displayed at the *bottom*, in correspondence of each session's name.

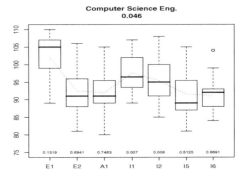

Fig. 12.7 Boxplot and results of the analysis of Electronic Engineering Course. The global *p*-values is shown on the *top* of the figure, the partial *p*-values are displayed at the *bottom*, in correspondence of each session's name.

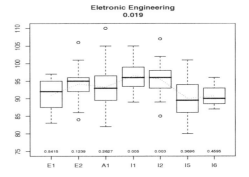

The global *p*-value of the Electronic Engineering course (Fig. 12.7) is moderately significant ($\Pi^G = 0.019$). By looking at the partial *p*-values of each group, we find out that session I2 has the minimum partial *p*-value equal to 0.003.

In the Management Engineering course (Fig. 12.8), there seems to be a peak group in E5 session. But there only two students in this session, and the sample

Fig. 12.8 Boxplot and results of the analysis of Management Engineering Course. The global *p*-values is shown on the *top* of the figure, the partial *p*-values are displayed at the *bottom*, in correspondence of each session's name.

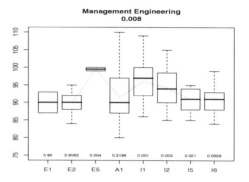

mean of this group is equal to 99.5. The global *p*-value is $\Pi^G = 0.008$, that is highly significant in favor of the umbrella alternative. The group with minimum partial *p*-values is again I1, with a *p*-value equal to 0.001.

Fig. 12.9 Boxplot and results of the analysis of Mechanical Engineering Course. The global *p*-values is shown on the *top* of the figure, the partial *p*-values are displayed at the *bottom*, in correspondence of each session's name.

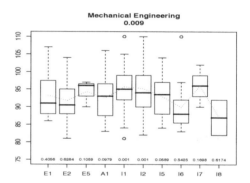

Finally, the evaluation of the Mechanical Engineering course (Fig. 12.9) does not show a clear trend. Nevertheless, the global *p*-value is significant at 1% level, and there is an *ex-aequo* in estimating the peak group. Groups I1 and I2 have a partial *p*-value equal to 0.001.

Chapter 13
Nonparametric methods for measuring concordance between rankings: a case study on the evaluation of professional profiles of municipal directors

Rosa Arboretti Giancristofaro, Mario Bolzan and Livio Corain

13.1 Introduction

The various areas of public administration management have, for several years, been the subject of considerable review, including legislative review. The particular importance of the need to redefine identities and responsibilities, as well as the knowledge and skills of the personnel operating in the field, has been recognised.

In this context a study was set up in association with the National Association of the Communes of Italy (Veneto section) which aimed to reconceptualise the municipal director profile. The study involved an initial stage in which privileged witnesses were interviewed through a Delphi survey in order to chart the dimensions constituting the municipal director's profile. Based on the dimensions identified, a questionnaire was prepared and presented to a sample of municipal directors from the communes of the Veneto. The questionnaire was split into two parts: firstly, the director was asked to assess the importance of possessing the qualities indicated by each item; secondly the director was asked to express an opinion in relation to the usefulness of investing in each of the dimensions. With regard to both parts of the questionnaire, it is of interest to obtain a ranking of the items. A method for constructing preference rankings based on the nonparametric combination procedure has been proposed to compete with the usual method based on the arithmetic mean. Subsequently in order to verify to what extent the rankings related to the two parts

---

Rosa Arboretti Giancristofaro
Department of Territory and Agro-Forestal Systems, University of Padua, Viale dell'Univerìstà 16, 35020 Legnaro (PD), Italy, e-mail: rosa.arboretti@unipd.it

Mario Bolzan
Department of Statistics, University of Padua, Via Cesare Battisti 241, 35121- Padua, Italy, e-mail: mbolzan@stat.unpd.it

Livio Corain
Department of Management and Engineering, University of Padua, Stradella S. Nicola 4, 36100 - Vicenza, Italy, e-mail: livio.corain@unipd.it

of questionnaire concord, a new permutation test for the evaluation of concordance between dependent rankings has been applied.

13.2 Sample survey of municipal directors' professional profiles

13.2.1 Context of the evaluation of the role of Communes and of municipal directors

The various areas of public administration management (role, services provided, professional profiles, material resources) have, for several years, been the subject of considerable review, including legislative review. The particular importance, in this process, of the need to redefine identities and responsibilities, as well as the knowledge and skills of the personnel operating in the field, has been recognised.

In particular, the coming into force of the Bassanini Law (59/1997 as amended) sparked a reorganisation process in all communes, diversified by speed, sharing and quality of results. These processes of change have led and continue to lead communes to reinvent themselves in terms of objectives, and from there the way in which they act.

In this process cultural dimensions are the most interesting. Revision of the role of clerk of the council, increase in the responsibilities of directors, promotion of subsidization, but above all division of roles between the political field and the operative field are the dimensions that mark the change.

The new national legislative context in fact requires of Local Authority functionaries and more generally of Public Authorities a training background aimed at changing perspectives: from an administration based on authority to one that provides citizens/area promoters with services, in which the functionary must be suitably trained to take on this role of not just management but also government collaborator with regard to the changes and implementation of the programmes.

The starting point as well as reference point is the typical professional figure described in the ministerial decree of 4 August 2000 which describes the employment prospects of graduates in the specific Public Authority disciplines of interest.

The "savoir-faire" required of senior degree-holding functionaries includes the ability to "effectively interpret change and organisational innovation in public and private organisations", to "assist public institutions and business, services and third sector private organisations in planning and implementing initiatives aimed at promoting the economic, social and civil development of communities", to "implement specific public policies" and "contribute to the management of human resources and trade-union relations".

The predicted impact of this change is on the one hand a repositioning of the role of local authority, on the other the repositioning of the role of municipal management.

With regard to the repositioning of municipal directors, the role's evolution places directors in direct contact with a variety of internal and external stakeholders, politicians or technicians with regard to whom it is necessary to reflect on the competencies and ability to act/interact. By stakeholders we mean all those with an interest in the activity of the Local Authority because they can influence activities, decisions, or can be influenced by them: mayor, councillors, citizens (organised into various levels), directors of an equal level, functionaries, municipalities, legislators, directors of other communes, directors of higher institutions (province, region, state).

Against this backdrop, initiatives aimed at reinterpreting and redesigning the professional profile of municipal directors are certainly desirable in order to consolidate the acquired knowledge and identify routes for development.

13.2.2 Sample survey among Communes of the Veneto

In 2008 a study was completed in association with the National Association of the Communes of Italy (Veneto section) which aimed to verify the perception and expectations of the role of the Commune with regard to the local area and community and to reconceptualise the municipal director profile (Bolzan et al., 2008).

The study provided for an initial stage in which privileged witnesses were interviewed through a Delphi survey in order to chart the dimensions that constitute the municipal director's profile in relation to his or her "savoir-faire" and "savoir-être". Based on 26 dimensions identified during this first stage, a questionnaire was then prepared and presented via e-mail (backed up by telephone calls) to a sample of municipal directors and clerks of the 581 communes of the Veneto.

The questionnaire was divided into two parts: firstly, respondents were asked to assess the importance of possessing the qualities indicated by the 26 items on a scale of 1 (of little importance) to 10 (very important); secondly, using the same scale, respondents were asked to express their opinion in relation to the usefulness of investing in each of the 26 dimensions in order to improve his or her professional profile.

The sample was made up of 193 communes from the 8 provinces of the Veneto, of which 47% had less than 5,000 inhabitants, 30% had between 5,000 and 10,000 inhabitants and the remaining 23% had over 10,000 inhabitants.

83% of the interviewed municipal directors and clerks (67% males) had a degree. 81% of clerks of the council had held their post for over 8 years (3% of respondents for less than 3 years).

13.3 Analysis of concordance between rankings

13.3.1 The construction of ranks

This paragraph discusses a method for constructing preference or importance rankings using evaluations taken from a sample of statistical units and expressed generally by scores in relation to a series of aspects representing partial dimensions of a given phenomenon of interest. It should be noted that the arithmetic mean (whether weighted or not) is the method mainly used for pooling preference ratings. A method based on the nonparametric combination of rankings has been proposed to compete with the usual method based on the arithmetic mean (Arboretti et al., 2005). A simulation study showed that this method generally performs better than the arithmetic mean.

Let us consider n subjects who are asked to rate each of M dimensions on a scale of 1–10. The problem of how to obtain this ranking, i.e. how to pool subject preferences, is addressed. Let X_{mi} be the rate of dimension m given by subject i, $i=1, ..., n$. Assume that if $X_{mi} > X_{m'i}$, then subject i rates dimension m better than dimension m'. In the literature this problem is usually solved by averaging subject ratings $\bar{X} = \sum_{i=1}^{n} X_{mi}/n$, $m=1, ..., M$, and dimension $\tilde{m}$ such that $\bar{X}_{\tilde{m}} = \max(\bar{X}_1, ..., \bar{X}_M)$ is then the best dimension with first rank position, dimension $\hat{m}$ such that $\bar{X}_{\hat{m}} = \max_{\{i=1,...,M, i \neq \tilde{m}\}}(\bar{X}_i)$ is the dimension with the second rank position, and so on. For simplicity's sake, it is assumed that there are no ties in ranking positions.

An alternative way to pool preferences is based on the nonparametric combination (NPC) ranking method. The procedure consists of three steps. In the first step a score for item m is computed as follows: $\eta_{mi} = \frac{\#(X_{mi} \geq X_{m'i}) + 0.5}{M+1}$, where $\#(X_{mi} \geq X_{m'i})$ indicates the rank transformation of X_{mi}. This step is repeated for each subject i and item m. With regard to relative rank transformation $\frac{\#(X_{mi} \geq X_{m'i})}{M}$ of X_{mi}, 0.5 and 1 have been added respectively to the numerator and the denominator to obtain η_{mi} varying in the open interval (0, 1). The reason for such corrections is merely computational, in order to avoid numerical problems with logarithmic transformations later on. Note that the scores η_{mi} are one-to-one increasingly related to the ranks $\#(X_{mi} \geq X_{m'i})$. By considering η_{mi}s after the first step, it is straightforward to obtain a (partial) ranking of the M dimensions for each subject, but it is the global dimension rank that is of interest. In the second step, the scores that subjects have assigned to dimension m are combined as follows: $C_m = -\sum_{i=1}^{n} \log(1 - \eta_{mi})$. This step is repeated for the remaining M-1 dimensions and a nonparametric combination of subjects' scores is performed. In the last step, the (global) ranking for item m is computed as $R_m = \#(C_m \geq C_{m'})$. Of course dimension $\tilde{m}$ with $R_{\tilde{m}} = M$ is the first rank position item, $\hat{m}$ with $R_{\hat{m}} = M - 1$ is the second, and so on. It should be noted that Fisher's omnibus combining function is used in the second step. Of course, other combining functions may be of interest in this problem (e.g. Liptak's function, Tippett's function, ...). The combining real function ϕ is chosen from class Φ of combining functions satisfying the following minimal properties:

1. ϕ must be continuous in all η_{mi} arguments, in that small variations in any subset of arguments imply small variations in the ϕ-index;
2. ϕ must be monotone non-decreasing in respect of each η_{mi} argument:

$$\phi(\ldots,\eta_{mi},\ldots) \geq \phi\left(\ldots,\eta'_{mi_i},\ldots\right) \text{ if } 1 > \eta_{mi} > \eta'_{mi} > 0, \, i=1,\ldots,n;$$

3. ϕ must be symmetric with respect to permutations of the arguments, in that if for instance $u_1,\ldots,u_n$ is any permutation of $1,\ldots,n$ then:

$$\phi\left(\eta_{u_1},\ldots,\eta_{u_n}\right) = \phi\left(\eta_1,\ldots,\eta_n\right).$$

Property 1 is obvious; property 2 means that if for instance two subjects have exactly the same values for all ηs, except for the m-th, then the one with $\eta_m > \eta'_m$ must have assigned at least the same satisfaction ϕ-index. Property 3 states that any combining function ϕ must be invariant with respect to the order in which subjects are processed.

It should also be noted that a central feature of NPC ranking is the possibility of assigning different degrees of importance to different types of subjects. If there is more interest in a certain group of subjects, we can assign them a weight of $0.5 < w < 1$. This weighted approach is taken into account in step two of the procedure by computing: $C_m = -\sum_{i=1}^{n} w_i \times \log(1 - \eta_{mi})$.

13.3.2 Hypothesis testing on concordance between rankings

The construction of several rankings of preference or importance, each referring to a set of dimensions or items indicative of a complex phenomenon, can lead to another objective, i.e. analysing the possible correlation between the obtained rankings (e.g. with the definition of the municipal director's professional profile, consider the ranking of the importance of possessing a certain set of competencies and the ranking of the need to invest to improve such competencies). Where stratification variables are present (e.g. size of the commune), analysis of the correlation could also be carried out both for each stratum and globally on the set of strata.

In the field of nonparametric combination of dependent permutation tests methodology (NPC, Pesarin, 2001), it is possible to find exact nonparametric solutions for these types of problems. The NPC approach can be considered a general methodology for many multivariate situations, such as cases in which sample sizes are lower than the number of observed variables or where there is missing data (even in the case of non-random missing data), or when some variables are categorical (ordinal or nominal) and others are quantitative and in many other complex situations. Additionally, the NPC solutions are of particular interest when there are restricted alternatives because in this specific case they prove to be particularly efficient in terms of power, easy to justify, simple to plan and, with the processors and software currently available, easy to carry out.

Suppose we have two rankings each made up of M items and we have a stratification variable (should there be more than one stratification variable, it is possible to make use of solutions such as *propensity score* in order to restore the situation to one stratification variable). The aim is to construct a test to verify the hypothesis of absence of correlation between the two rankings both globally on all the strata, and singularly by stratum. The problem can be solved by first considering a set of M partial permutation tests, each directed at testing the significance of the correlation between corresponding pairs of items in the rankings, followed by their nonparametric combination into a global test.

For ordered variables, a suitable test statistic for permutation partial tests is based on Spearman's rank-order correlation coefficient (a reference is provided in Siegel and Castellan, 1988). This measure of association capture in a single number (varying from -1 to 1) the relationship between two ordered data series. In particular we can use this measure of association if we are reluctant to make the assumption of bivariate normality in relation to the two data series.

Spearman's rank-order correlation coefficient uses ranks derived from the raw data. Specifically, if the data are represented in the form of a $r \times c$ contingency table, formed from n observations cross-classified into r row categories and c column categories with x_{ij} of the observations falling into row-category i and column-category j, and $m_1, m_2, \ldots, m_r$, and $n_1, n_2, \ldots, n_c$ are respectively the marginal row totals and column totals, Spearman's measure uses $u_i = m_1 + m_2 + \ldots + m_i - 1 + (m_i+1)/2$, for $i = 1, 2, \ldots, r$, and $v_j = n_1 + n_2 + \ldots + n_j - 1 + (n_j+1)/2$, for $j = 1, 2, \ldots, c$, to obtain:

$$S = \frac{\sum_{i=1}^{r} \sum_{j=1}^{c} x_{ij} (u_i - \bar{u})(v_i - \bar{v})}{\sqrt{\sum_{i=1}^{r} m_i (u_i - \bar{u})^2} \sqrt{\sum_{j=1}^{c} n_j (v_i - \bar{v})^2}}$$

where:

$$\bar{u} = \sum_{i=1}^{r} m_i u_i / n, \bar{v} = \sum_{j=1}^{c} n_j v_j / n.$$

When implementing permutation partial tests, we have the problem of determining the permutation cumulative distribution function associated to each test. The exact derivation of the permutation cumulative distribution function connected to a statistical test under the null hypothesis is computationally difficult in cases where sample sizes are not very small. Hence, the exact derivation of the permutation distribution associated with any statistic of interest is at the very least impractical, if not impossible. The problem may be dealt with by means of Monte Carlo simulation from permutational space. This solution leads to a resampling technique, conditional on the pooled data set. This is done through a without-replacement resampling procedure, also called the CMC (Conditional Monte Carlo) method (Pesarin, 2001). The CMC method on the pooled data set is a random simulation of all possible permutations of the same data under H_0. Hence, in order to obtain an estimate of the permutation distribution under H_0 of the test statistic, and therefore proceed with the calculation of the *p*-value $\lambda_m, m = 1,...,M$ associated with each partial test, a CMC procedure can be used.

For the sake of simplicity, let $X_m, m = 1,...,M$, indicate either a continuous or a categorical response variable. Let T_m be the statistic related to the partial test of the *m*-th variable.

The CMC procedure considers the following steps:

1. calculate the *M*-dimensional vector of the test statistics, each one related to partial tests for observed data X:

$$T_0 \atop M \times 1 = T(X) = [T_{m0} = T_m(X), m = 1,...,M];$$

2. consider a data permutation X^* by a random resampling of X, in order to randomly assign every individual data vector to a proper group and then calculate the value of the test statistic $T^* = T(X^*)$;
3. independently repeat the above step 2. B times. $\{T_r^*, r = 1,...,B\}$ indicates the resulting vectors of B conditional CMC-iterations;
4. the empirical *M*-variate cumulative distribution function (EDF) is given by:

$$\hat{F}_B(z|X) = \left[\frac{1}{2} + \sum_{r=1}^{B} I(T_r^* \leq z)\right] \Big/ (B+1), \forall z \in R^M$$

where $\mathbf{I}(.)$ is the indicating function. $\hat{F}_B(z|X)$ gives an estimate of corresponding permutation *M*-dimensional distributions $F(z|X)$ of T, and the functions:

$$\hat{L}_i(z|X) = \left[\frac{1}{2} + \sum_{r=1}^{B} I(T_{ir}^* \geq z)\right] \Big/ (B+1), m = 1,...,M,$$

give an estimate $\forall z \in R^1$ of permutation marginal significance levels $L_m(z|X) = \Pr\{T_m^* \geq z|X\}$. Consequently:

$$\hat{\lambda}_m = \hat{L}_m(T_{m0}|X) = \left[\frac{1}{2} + \sum_{r=1}^{B} I(T_{mr}^* \geq T_{m0})\right] \Big/ (B+1), m = 1,...,M$$

give an estimate of marginal *p*-values $\lambda_m = \Pr\{T_m^* \geq T_{m0}|X\}$, relative to partial tests T_m.

If $\hat{\lambda}_m \leq \alpha$, the null hypothesis relating to the *m*-th variable is rejected at significance level α.

Note that in comparison with standard estimators of the empirical cumulative distribution function, values 1/2 and 1 were added respectively to the numerator and denominator to obtain estimates of *p*-values in the open interval (0, 1). This does not change inferential conclusions, does not influence the asymptotic behavior of estimators and proves to be useful for the combination of partial tests.

The nonparametric combination of M partial tests allows us to obtain a suitable solution for the global hypothesis system. The nonparametric combination of

M p-values associated with the partial tests into the global test is achieved by using an appropriate real, continuous, non-increasing and non degenerate function $\psi : (0,1) \to R^1$. Let us assume the combining function ψ has the following features:

a. it is monotone non-increasing in respect of each argument: $\psi(\ldots,\lambda_m,\ldots) > \psi(\ldots,\lambda'_m,\ldots)$ if $\lambda_m < \lambda'_m$, $m = 1,\ldots,M$,
b. it attains its upper limit $\bar{\psi}$, possibly not finite, when at least one argument goes to 0: $\psi(\ldots,\lambda_m,\ldots) \to \bar{\psi}$ for $\lambda_m \to 0$, and the negative lower limit $\underline{\psi}$ when at least one argument goes to 1: $\psi(\ldots,\lambda_m,\ldots) \to \underline{\psi}$ for $\lambda_m \to 1$;
c. $\forall \alpha > 0$, the acceptance region is limited: $\underline{\psi} < T''_{\alpha/2} < T'' < T''_{(1-\alpha/2)} < \bar{\psi}$.

The above properties define a class Ψ of combining functions (Pesarin, 2001). Some of the functions most often used to combine independent tests (Fisher, Lancaster, Liptak, Tippett, Mahalanobis, etc.) are included in this class. Let us look at Tippett combinig function which is given by $T''_T = \max_{1 \leq m \leq M}(1 - \lambda_m)$

By again using the results of the same conditional simulation procedure CMC, used to estimate the p-values of partial tests, we achieve the nonparametric combination of M partial tests T_m, $m = 1, \ldots, M$. Let us start again from step (4), the NPC considers the following steps:

5. we determine the combined observed value of the global test using the same CMC results as the previous step, given by:

$$T''_0 = \psi\left(\hat{\lambda}_1, \ldots, \hat{\lambda}_M\right);$$

6. the r-th combined value is then calculated by:

$$T''_r* = \psi(\lambda_{1r}*, \ldots, \lambda_{Mr}*), r = 1, \ldots, B;$$

7. estimate the p-value for the combined test T'' in the following way:

$$\hat{\lambda}''_\psi = \sum_{r=1}^{B} I\left(T''_r* \geq T''_0\right) \bigg/ B,$$

8. if $\hat{\lambda}''_\psi \leq \alpha$, we reject the null hypothesis H_0 at significance level α.

When a stratification variable is present, the nonparametric combination takes place firstly within each stratum and then between the strata.

13.3.3 Closed testing procedure

Multiple comparisons and multiple testing problems arise frequently in statistical data analysis, and it is important to address them appropriately. Actually, the problem of multiplicity control arises in all cases where the number of hypotheses to be tested is greater than one. Such partial tests, possibly after adjustment for multiplicity (Westfall et al., 1999), may be useful for marginal or separate inferences. If they are jointly considered they provide information on a general overall or global hypothesis, which typically represents the true objective of the majority of multivariate testing problems. In order to produce a valid test for the combination of a large number of p-values, we must guarantee that such test is unbiased and produces, therefore, p-values below the significance level with a probability less or equal to α itself. This combination could be very troublesome unless we are working in a permutation framework. A Bonferroni correction is valid but the conservativeness of this solution is often unacceptable for both theoretical and practical purposes. Actually, this combination loses power in case of dependence between p-values. On the contrary, using appropriate permutation methods, dependencies may be controlled. With reference to multiple testing procedures mentioned before, these have their starting point in an overall test and look for significant tests on partial contrasts. Conversely combination procedures start with a *a set of partial tests*, each appropriate for a partial aspect, and look for joint analyses leading to global inferences. The global p-value obtained through NPC procedure of p-values associated to sub-hypotheses is an exact test, thus providing a weak control of the multiplicity. The inference in this case must be limited to the global evaluation of the phenomenon. Due to the use of NPC methods, a more detailed analysis may be carried out. Actually, what is important is to select potentially active hypotheses (i.e. under the alternative). A correction of each single p-value is hence necessary in this case. A possible solution within a nonparametric permutation framework is represented by Closed testing procedures (Westfall et al., 1999). A property that is generally required is the strong control of the Familywise Error Rate (FWE), i.e. the probability of making one or more errors on the whole of the considered hypotheses (Marcus et al., 1976). On the other hand, a weak control of the FWE means simply controlling α for the global test (i.e. the test where all hypotheses are null). Although the latter is a more lenient control, it does not allow the selection of active variables because it simply produces a global p-value that does not allow interesting hypotheses to be selected, so the former is usually preferred because it makes inference on each (univariate) hypothesis (Finos and Salmaso, 2007). An alternative approach to multiplicity control is given by the False Discovery Rate (FDR). This is the maximum proportion of type I errors in the set of elementary hypotheses. The FWE guarantees a more severe control than the FDR, which in fact only controls the FWE in the case of global null hypotheses, i.e. when all involved hypotheses are under H_0 (Benjamini and Hochberg, 1995). In confirmatory studies, for example, it is usually better to strongly control the FWE, thus ensuring an adequate inference when you want to avoid making even one error. On the contrary, when it is of interest to highlight a pattern of potentially involved variables, especially when dealing with thousands of variables, the FDR would appear to be a more reasonable approach. In this way it is accepted that part (no greater than the α proportion) of the rejected hypotheses are in fact under the null (Finos and Salmaso, 2007).

The goal of multiple testing procedures is to control the "maximum overall Type I error rate", i.e. the maximum probability that one or more null hypotheses is rejected incorrectly. This quantity also goes by the name "Maximum Experimentwise Error Rate" (MEER).

With reference to the closed testing, here we give just some hints and we refer the reader to Westfall et al. (1999). Suppose we wish to test hypotheses H_1, H_2, H_3 and H_4, e.g. concerning 4 variables. Hence, with reference to the Fig. 13.1 we start applying closed testing. The closed testing method works as follows.

1. Test each hypothesis H_1, H_2, H_3 and H_4 using an appropriate α-level test.
2. Create the "closure" of the set, which is the set of all possible intersections among H_1, H_2, H_3 and H_4 (in this case the hypotheses H_{12}, H_{13}, H_{14}, H_{23}, H_{24}, H_{34}, H_{123}, H_{134}, H_{234} and H_{1234}). In Fig. 13.1 we illustrate the procedure. We have enumerated all the possible intersections, but of course we are interested only in some of these intersections. Actually some of these are useful for inferential purpose, some other are only instrumental and are not investigated. Intersections of interest are represented by the red bounded boxes, corresponding respectively to the variable level (i.e., H_1, H_2, H_3 and H_4), to the domain level (i.e., H_{12} and H_{34}) and to the global test (H_{1234}).
3. Test each intersection using an appropriate α-level test. In general any test that is valid for the given intersection.
4. You may reject any hypothesis H_i, with control of the MEER, when the following conditions both hold

   – The test of H_i itself yields a statistically significant result, and
   – The test of every intersection hypothesis that includes H_i is statistically significant.

Hence, a statistically significant result has been obtained for the H_3 test, as well as a significant result for all hypotheses that include H_3, in this case, H_{13}, H_{23}, H_{34}, H_{123}, H_{134}, H_{234} and H_{1234} (blue boxes in Fig. 13.1). Since the p-value for one of the including tests, the H_{1234} test in this case, is greater than 0.05, you may not reject the H_3 test at the MEER = 0.05 level. In this example, we could reject the H_3 hypothesis for MEER levels as low as, but no lower than 0.0618, since this is the largest p-value among all hypotheses containing H_3. This suggests an informative way of reporting the results of a closed testing procedure. When using a closed testing procedure, the adjusted p-value for a given hypothesis H_i is the maximum of all p-values for tests that include H_i as a special case (including the p-value for the H_i test itself). The adjusted p-value for testing H_3 is, therefore, formally computed as $\max(0.0067, 0.0220, 0.0285, 0.0285, 0.0570, 0.0580, 0.0600, 0.0618) = 0.0618$.

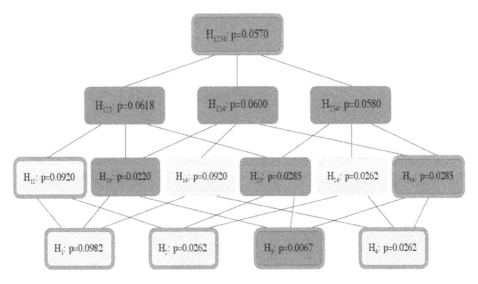

Fig. 13.1 Illustration of the closed testing procedure ($k = 4$).

13.3.4 Rankings of the municipal director's professional profile in the Communes of the Veneto survey

In relation to the two parts of the Communes of the Veneto questionnaire (the first part in which respondents were asked to assess the importance of possessing the qualities indicated by the items; the second in which they was asked to express their opinion in relation to the usefulness of investing in each of the dimensions), it is of interest to obtain a ranking of the items and subsequently verify to what extent the two rankings concord. This verification is carried out by applying the procedure discussed in the previous paragraph.

Table 13.1 shows the rankings of the importance and need for development of the various competencies.

Table 13.2 shows the correlation indices for each competence, calculated by considering the positions assigned in the two rankings. The different colours of the table's cells correspond to the significance of the partial tests for each competence.

Tables 13.3, 13.2, 13.3, 13.4 and 13.5 show the results of the analyses carried out taking as stratification variable the commune size. The p-values relating to combinations by stratum and to the global combination are illustrated in Table 13.6, where p-values have been adjusted for multiplicity using a closed testing procedure. P-values are displayed according to a grey scale sequence:

| p-values | < 0.001 | < 0.001 | < 0.05 | n.s. |

Table 13.1 Rankings of the importance of competencies and of the need for development of competencies

	Importance of competencies			Need for development of competencies	
1	Ability to motivate staff	1.0000	1	Ability to motivate staff	1.0000
2	Ability to build teams, integrate skills	0.9615	2	Ability to build teams, integrate skills	0.9615
3	Ability to make decisions	0.9231	3	Ability to obtain results	0.9231
4	A clear knowledge of the objectives of the local authority for which one works	0.8846	4	Ability to organise planning	0.8846
5	Loyal in relationships	0.8462	5	Ability to make decisions	0.8462
6	Ability to obtain results	0.8077	6	Ability to manage conflict	0.8077
7	Authoritative leader, not authoritarian	0.7692	7	Authoritative leader, not authoritarian	0.7692
8	A sense of duty	0.7308	8	A clear knowledge of the objectives of the local authority for which one works	0.7308
9	Ability to organise planning	0.6923	9	Being creative (open to innovation, ability to think up solutions)	0.6923
10	Ability to inspire trust	0.6538	10	Ability to interpret the local area (needs and resources)	0.6538
11	Knowledge of how administrative processes operate	0.6154	11	Ability to give reasons for choices	0.6154
12	Ability to manage conflict	0.5769	12	Ability to exploit the produced knowledge	0.5769
13	Being creative (open to innovation, ability to think up solution)	0.5385	13	Ability to communicate with local citizens	0.5385
14	Ability to give reasons for choices	0.5000	14	Able with regard to management control	0.5000
15	Technical know-how linked to the specificity of the role	0.4615	15	Ability to inspire trust	0.4615
16	Ability to evaluate situations case by case, not ideologically	0.4231	16	Technical know-how linked to the specificity of the role	0.4231
17	Ability to interpret the local area (needs and resources)	0.3846	17	Accountable	0.3846
18	Ability to exploit the produced knowledge	0.3462	18	Knowledge of how administrative processes operate	0.3462
19	Ability to communicate with local citizens	0.3077	19	A sense of duty	0.3077
20	Ability to relate to politicians	0.2692	20	Ability to evaluate situations case by case, not ideologically	0.2692
21	Accountable	0.2308	21	Loyal in relationships	0.2308
22	Autonomous and independent of political power	0.1923	22	A governing mentality	0.1923
23	Able with regard to management control	0.1538	23	Ability to relate to politicians	0.1538
24	A governing mentality	0.1154	24	Basic knowledge of cross-sector themes (e.g computers, statistics, quality)	0.1154
25	Basic knowledge of cross-sector themes (e.g. computers, statistics, quality)	0.0769	25	Autonomous and independent of political power	0.0769
26	An administrative mind	0.0385	26	An administrative mind	0.0385

13.3.5 Discussion

Analysis of the rankings for all interviewed communes (Table 13.1) and by size of the commune (Tables 13.3 and 13.4) highlights recognition common to both small and medium/large communes of the importance of competencies regarding the social dimension of the municipal director's profile. This profile should play a clear, collaborative role proposing solutions specific to relationships within the municipal

13 Nonparametric methods for measuring concordance between rankings

Table 13.2 Index of correlation between rankings by competence

	Correlation Index
Technical know-how linked to the specificity of the role	0.4493
Basic knowledge of cross-sector themes (e.g. computers, statistics, quality)	0.3693
Knowledge of how administrative processes operate	0.3157
A clear knowledge of the objectives of the local authority for which one works	0.2807
An administrative mind	0.4868
Ability to organise planning	0.4793
Able with regard to management control	0.4967
Accountable	0.4809
Ability to relate to politicians	0.3596
Ability to communicate with local citizens	0.4166
Ability to motivate staff	0.3028
Ability to build teams, integrate skills	0.3578
Ability to exploit the produced knowledge	0.4928
Ability to manage conflict	0.2867
Ability to make decisions	0.2827
A governing mentality	0.5269
Loyal in relationships	0.2478
Ability to inspire trust	0.2903
A sense of duty	0.1688
Autonomous and independent of political power	0.3076
Authoritative leader, not authoritarian	0.1166
Ability to obtain results	0.3893
Ability to give reasons for choices	0.3197
Ability to evaluate situations case by case, not ideologically	0.2637
Being creative (open to innovation, ability to think up solutions)	0.3548
Ability to interpret the local area (needs and resources)	0.4012

structure (e.g. ability to motivate staff, ability to build teams and integrate skills). The necessary, priority competencies can also be associated with the identity of the authority for which the director works (a clear knowledge of the objectives of one's local authority), which is expressed in terms of an ability to make decisions and look towards results, not merely dealing with administrative practices. Such competencies do however require directors to advance in a direction that is not of ordinary administration or, worse still, play it by ear. Being a director does not mean being authoritarian – it involves team playing and he or she must be credible and convincing. What count are results, not the management of what is already present.

Less characteristic of the director's role are the competencies or culture of an administrative or management control nature, and transversal knowledge of a specific or technical nature. Sitting at the bottom of the ranking, there is no recognition of a need for training inputs to support their further development. Also low in the ranking is the ability to relate with politicians and with the local community (except in large communes where greater importance is given to initiatives favouring communication with citizens).

Table 13.3 Ranking of the importance of competencies by size of commune

	Size of commune					
	< 5,000		5,000–10,000		≥ 10,000	
1	Ability to motivate staff	1.0000	Ability to motivate staff	1.0000	Ability to build teams, integrate skills	1.0000
2	Loyal in relationships	0.9615	Authoritative leader, not authoritarian	0.9615	Ability to motivate staff	0.9615
3	A clear knowledge of the objectives of the local authority for which one works	0.9231	Ability to make decisions	0.9231	Ability to make decisions	0.9231
4	Ability to build teams, integrate skills	0.8846	A clear knowledge of the objectives of the local authority for which one works	0.8846	A clear knowledge of the objectives of the local authority for which one works	0.8846
5	Ability to obtain results	0.8462	Loyal in relationships	0.8462	Loyal in relationships	0.8462
6	Ability to make decisions	0.8077	Ability to build teams, integrate skills	0.8077	Ability to inspire trust	0.8077
7	Authoritative leader, not authoritarian	0.7692	A sense of duty	0.7692	Ability to obtain results	0.7692
8	A sense of duty	0.7308	Ability to obtain results	0.7308	Ability to manage conflict	0.7308
9	Ability to inspire trust	0.6923	Ability to organise planning	0.6923	Authoritative leader, not authoritarian	0.6923
10	Ability to organise planning	0.6538	Knowledge of how administrative processes operate	0.6538	A sense of duty	0.6538
11	Knowledge of how administrative processes operate	0.6154	Being creative (open to innovation, ability to think up solutions)	0.6154	Ability to organise planning	0.6154
12	Ability to give reasons for choices	0.5769	Ability to inspire trust	0.5769	Ability to give reasons for choices	0.5769
13	Ability to manage conflict	0.5385	Technical know-how linked to the specificity of the role	0.5385	Being creative (open to innovation, ability to think up solutions)	0.5385
14	Being creative (open to innovation, ability to think up solutions)	0.5000	Ability to interpret the local area (needs and resources)	0.5000	Ability to communicate with local citizens	0.5000
15	Technical know-how linked to the specificity of the role	0.4615	Ability to evaluate situations case by case, not ideologically	0.4615	Ability to exploit the produced knowledge	0.4615
16	Ability to evaluate situations case by case, not ideologically	0.4231	Ability to give reasons for choices	0.4231	Accountable	0.4231
17	Ability to exploit the produced knowledge	0.3846	Autonomous and independent of political power	0.3846	Ability to evaluate situations case by case, not ideologically	0.3846
18	Ability to interpret the local area (needs and resources)	0.3462	Ability to communicate with local citizens	0.3462	Technical know-how linked to the specificity of the role	0.3462
19	Autonomous and independent of political power	0.3077	Ability to manage conflict	0.3077	Ability to interpret the local area (needs and resources)	0.3077
20	Ability to relate to politicians	0.2692	Ability to exploit the produced knowledge	0.2692	Ability to relate to politicians	0.2692
21	Ability to communicate with local citizens	0.2308	Able with regard to management control	0.2308	Knowledge of how administrative processes operate	0.2308
22	Accountable	0.1923	Ability to relate to politicians	0.1923	Able with regard to management control	0.1923
23	Able with regard to management control	0.1538	Accountable	0.1538	Autonomous and independent of political power	0.1538
24	A governing mentality	0.1154	A governing mentality	0.1154	A governing mentality	0.1154
25	Basic knowledge of cross-sector themes (e.g. computers, statistics, quality)	0.0769	Basic knowledge of cross-sector themes (e.g. computers, statistics, quality)	0.0769	Basic knowledge of cross-sector themes (e.g. computers, statistics, quality)	0.0769
26	An administrative mind	0.0385	An administrative mind	0.0385	An administrative mind	0.0385

13 Nonparametric methods for measuring concordance between rankings

Table 13.4 Ranking of the need for development of competencies by size of commune

	Size of commune					
	< 5,000		5,000–10,000		≥ 10,000	
1	Ability to motivate staff	1.0000	Ability to motivate staff	1.0000	Ability to obtain results	1.0000
2	Ability to obtain results	0.9615	Ability to organise planning	0.9615	Ability to motivate staff	0.9615
3	Ability to build teams, integrate skills	0.9231	Ability to build teams, integrate skills	0.9231	Ability to build teams, integrate skills	0.9231
4	Ability to organise planning	0.8846	Ability to make decisions	0.8846	Ability to make decisions	0.8846
5	Ability to manage conflict	0.8462	Able with regard to management control	0.8462	Ability to manage conflict	0.8462
6	Ability to make decisions	0.8077	Ability to manage conflict	0.8077	Ability to organise planning	0.8077
7	A clear knowledge of the objectives of the local authority for which one works	0.7692	Ability to obtain results	0.7692	Ability to inspire trust	0.7692
8	Authoritative leader, not authoritarian	0.7308	Authoritative leader, not authoritarian	0.7308	Ability to exploit the produced knowledge	0.7308
9	Being creative (open to innovation, ability to think up solutions)	0.6923	Ability to interpret the local area (needs and resources)	0.6923	Accountable	0.6923
10	Ability to give reasons for choices	0.6538	Being creative (open to innovation, ability to think up solutions)	0.6538	Authoritative leader, not authoritarian	0.6538
11	Ability to interpret the local area (needs and resources)	0.6154	A clear knowledge of the objectives of the local authority for which one works	0.6154	Being creative (open to innovation, ability to think up solutions)	0.6154
12	Ability to inspire trust	0.5769	Ability to communicate with local citizens	0.5769	Ability to interpret the local area (needs and resources)	0.5769
13	Ability to communicate with local citizens	0.5385	Ability to exploit the produced knowledge	0.5385	Able with regard to management control	0.5385
14	Ability to exploit the produced knowledge	0.5000	Ability to give reasons for choices	0.5000	A clear knowledge of the objectives of the local authority for which one works	0.5000
15	Knowledge of how administrative processes operate	0.4615	Technical know-how linked to the specificity of the role	0.4615	A sense of duty	0.4615
16	Technical know-how linked to the specificity of the role	0.4231	Knowledge of how administrative processes operate	0.4231	Ability to give reasons for choices	0.4231
17	A sense of duty	0.3846	Accountable	0.3846	Ability to communicate with local citizens	0.3846
18	Ability to evaluate, situations case by case not ideologically	0.3462	Basic knowledge of cross-sector themes (e.g. computers, statistics, quality)	0.3462	Loyal in relationships	0.3462
19	Able with regard to management control	0.3077	Ability to inspire trust	0.3077	Ability to evaluate situations case by case, not ideologicaly	0.3077
20	Accountable	0.2692	A governing mentality	0.2692	A governing mentality	0.2692
21	Ability to relate to politicians	0.2308	Loyal in relationships	0.2308	Technical know-how linked to the specificity of the role	0.2308
22	Loyal in relationships	0.1923	A sense of duty	0.1923	Knowledge of how administrative processes operate	0.1923
23	A governing mentality	0.1538	Ability to relate to politicians	0.1538	Basic knowledge of cross-sector themes (e.g. computers, statistics, quality)	0.1538
24	Autonomous and independent of political power	0.1154	Ability to evaluate situations case by case, not ideologically	0.1154	Autonomous and independent of political power	0.1154
25	Basic knowledge of cross-sector themes (e.g. computers, statistics, quality)	0.0769	Autonomous and independent of political power	0.0769	Ability to relate to politicians	0.0769
26	An administrative mind	0.0385	An administrative mind	0.0385	An administrative mind	0.0385

Table 13.5 Correlation index between rankings by competence and size of commune

	Correlation index		
	Size of commune		
	< 5,000	5,000–10,000	≥ 10,000
Technical know-how linked to the specificity of the role	0.4381	0.4014	0.5474
Basic knowledge of cross-sector themes (computers, statistics etc.)	0.4829	0.1818	0.4826
Knowledge of how administrative processes operate	0.5234	0.1468	0.0162
Knowledge of objectives of local authority for which one works	0.4389	0.2912	−0.0462
An administrative mind	0.4996	0.4250	0.4604
Ability to organise planning	0.5240	0.4855	0.4354
Able with regard to management control	0.5301	0.4154	0.5580
Accountable	0.4658	0.4763	0.4684
Ability to relate to politicians	0.5142	0.1331	0.2422
Ability to communicate with local citizens	0.5160	0.3214	0.2137
Ability to motivate staff	0.4517	0.1074	0.1654
Ability to build teams, integrate skills	0.4292	0.3529	0.0380
Ability to exploit the produced knowledge	0.5145	0.6499	0.1636
Ability to manage conflict	0.4154	0.1928	−0.0538
Ability to make decisions	0.3854	0.3914	−0.0093
A governing mentality	0.5804	0.5534	0.3102
Loyal in relationships	0.3048	0.0738	0.1179
Ability to inspire trust	0.3532	0.2505	0.2192
A sense of duty	0.2537	−0.1290	0.3532
Autonomous and independent of political power	0.4046	0.0497	0.3071
Authoritative leader, not authoritarian	0.1897	−0.0165	0.1743
Ability to obtain results	0.4278	0.3651	0.2921
Ability to give reasons for choices	0.4243	0.2651	−0.0019
Ability to evaluate situations case by case, not ideologically	0.4321	0.0509	0.1083
Being creative (open to innovation, ability to think up solutions)	0.4403	0.1117	0.4950
Ability to interpret the local area (needs and resources)	0.4605	0.4168	0.3701

Table 13.6 Significance of the correlation between rankings by stratum (size of commune) and global

Size of the Commune	p-value
< 5,000	0.000999
5,000–10,000	0.008991
≥ 10,000	0.000999
Global	0.000999

Having identified this ordering, it is natural to consider the distance between this profile and the one currently present, particularly the profile that the clerks of the council or directors, as central figures in the "administrative machine", believe they have, or better still which aspects they believe they need to develop. Analysis of the rankings dealing with the need to develop the considered competencies (Tables 13.1 and 13.4) suggests a consideration that appears to be symmetric to the interpretation of the importance of the competencies. The greatest need for training is identified in specific operational dimensions (ability to obtain results, plan, integrate skills, in short, ability to organise and manage resources). With these results it would seem that the culture of responsibility and transparency has begun to spread to all levels of Italian administration, including suburban local authorities where the elective office – starting with the mayor – is "measured" on the basis of "results" and no longer, or at least not exclusively as in the past, on the basis of political persuasion. Heads of administrations are thus motivated to choose staff who pursue visible objectives that can be presented to the people. Less need is felt for aspects that are perhaps already common within the public authority environment, and thus considered obsolete. It is perhaps felt that the new local authority operating framework requires greater commitment to investing in sectors the command of which makes and will make the difference in terms of quality of the profession, sectors in which until now traditional education, both university and postgraduate, have invested insufficiently.

It can be seen that the dimensions to which greater importance is given are also those that require greater investment, therefore indicating a very clear perspective. Substantially, high opinions of the importance of single competencies correspond with high opinions of the need to develop such competencies (Tables 13.2 and 13.5); the global correlation between the two rankings is significant, both for the total number of interviews ($p < 0.0001$) and for the three dimensions of the communes (Table 13.6). Analysis of the correlation between necessary competencies and the respective need to invest in training provides interesting points for reflection, particularly if the analysis is carried out separately by size of the commune to which the interviewee belongs (Table 13.5). The results show that in smaller communes all correlations are significant (at times with elevated significance) and confirm that the identity perceived as necessary for directors requires adequate wide-ranging investment, perhaps in the knowledge that these local authorities are the ones that will play a propulsive role in the area's development, as well as offering an excellent opportunity for directors to invest in their own professional future. Larger communes, on the other hand, display a more articulate panorama: a less consistent spread of significant correlations perhaps indicates greater caution towards what should be invested in, or greater experience or professional maturity that leads to the association of certain competencies with specific educators or training agencies which cannot always – when contacted – live up to expectations. In particular it can be seen that among the non-significant correlations for medium and large communes, seven relate to the same competencies while the remainder relate to different competencies. The need for competencies would appear not to be correlated to the respective importance of training in the following cases: knowledge of how administrative processes operate, ability to motivate staff, ability to manage conflict, loyal in relationships, ability to inspire trust, authoritative leader and not authoritarian, ability to evaluate situations case by case and not ideologically. One possible interpretation is that these competencies are the expression of abilities which are "not learnt at school" but are the result of years of experience, particularly working in smaller communes where one can reasonably expect to start and where the need to learn what appears to be necessary for the profession is more greatly felt (therefore the correlations are all significant).

In short, analysis seems to highlight a recognition of the complexity of the municipal director's role with regard to the new prospects of local authority positioning, and therefore the need to affirm a new identity for their professional profile that goes beyond the abilities traditionally associated with it and probably already acquired. As a result, therefore, new forms of training are required, especially in relation to the ability to act/interact with various interlocutors both within and outwith the administration.

As regards theoretical aspects, the above presented method of analysis is very flexible and allows the experimenter to perform hypothesis testing on concordance between dependent rankings for problems with ordinal data. Where stratification variables are present, the nonparametric combination procedure along with closed testing allows us to obtain a suitable solution for the global hypothesis system for each stratum of stratification variables and for each dimension belonging to the overall ranking.

References

Aber MS, McArdle JJ (1991) Latent growth curve approaches to modeling the development of competence. In: Chandler, M. and Chapman, M. (Eds.), Criteria for Competence Hillsdale, Lawrence Earlbaum Associates I: NJ, pp. 231–258

Adams AJ, Wilson M, Wang WC (1997) The multidimensional random coefficients multinomial logit model. Applied Psychological Measurement 21(1):1–23

Afshartous D, Wolf M (2007) Avoiding 'data snooping' in multilevel and mixed effects models. Journal of the Royal Statistical Society A 170:1035–1059

Agresti A (2002) Categorical Data Analysis. 2nd edn. John Wiley and Sons, New York

Agresti A, Yang MC (1987) An empirical investigation of some effects of sparseness in contingency tables. Computational Statistics and Data Analysis 5:9–21

Aiello F, Attanasio M (2004) How to transform a batch of simple indicators to make up a unique one? In: Atti della XLII Riunione Scientifica della Società Italiana di Statistica, Sessioni Plenarie e Specializzate, CLEUP, Padova, pp. 327–338

AISE (2009) A.I.S.E. detergent test protocol – 2009. WWW page, URL www.aise.eu/aisedetergenttestprotocol2009/

Aitkin M (1999) A general maximum likelihood analysis of variance components in generalized linear models. Biometrics 55:117–128

Aitkin M, Longford N (1986) Statistical modelling issues in school effectiveness studies. Journal of the Royal Statistical Society A 149:1–43

Albert JH (1992) Bayesian estimation of normal ogive item response curves using Gibbs sampling. Journal of Educational Statistics 17(3):251–269

Almalaurea (2006) Condizione occupazionale dei laureati. Consorzio Interuniversitario Almalaurea, Bologna

Alvarez-Farizo B, Hanley N (2002) Using conjoint analysis to quantify public preferences over the environmental impact of wind farms. An example from spain. Energy Policy 30:107–116.

Anscombe FJ (1960) Rejection of outliers. Technometrics 2:123–147

Arabie P, Hubert LJ (1990) The bond energy algorithm revisited. IEEE Transactions on Systems, Man and Cybernetics 20:268–274

Arboretti GR, Marozzi M, Salmaso L (2005) Nonparametric pooling and testing of preference ratings for full-profile conjoint analysis experiments. Journal of Modern Applied Statistical Methods 4:545–552

Arboretti GR, Basso D, Bonnini S, Corain L (2008) A robust approach for treatment ranking within the multivariate one-way anova layout. In: Proceedings in Computational Statistics (Edited by Paula Brito), International Conference on Computational Statistics, Porto – Portugal, August 24th–29th 2008, pp. 649–657

Atkinson AC, Riani M (2000) Robust Diagnostic Regression Analysis. Springer, New York

Atkinson AC, Riani M, Cerioli A (2004) Exploring Multivariate Data with the Forward Search. Springer, New York

Baker GA (1954) Factor analysis of relative growth. Growth 18:137–143.

Balirano G, Corduas M (2008) Detecting semiotically expressed humor in diasporic tv productions. HUMOR: International Journal of Humor Research 3:227–251

Barnett V (1976) The ordering of multivariate data. Journal of the Royal Statistical Society, RSS – Series A 139:318–339

Barnett V (1988) Outlier and order statistics. Communications in Statistics Theory and Methods 17:2109–2118

Barnett V, Lewis T (1993) Outliers in Statistical Data, 3rd edn. John Wiley and Sons, New York.

Barone S, Lombardo A, Tarantino P (2007) A weighted logistic regression for conjoint analysis and kansei engineering. Quality Reliability Engineering International 23:689–706

Bartholomew DJ (1995) Spearman and the origin and development of factor analysis. British Journal of Statistical and Mathematical Psychology 48:211–220

Bartholomew DJ, Knott M (1999) Latent Variable Models and Factor Analysis. Hodder Arnold, London

Bartholomew DJ, Tzamourani P (1999) The goodness of fit of latent trait models in attitude measurement. Sociological Methods and Research 27:525–546

Basso D, Pesarin F, Salmaso L, Solari A (2009) Permutation Tests for Stochastic Ordering and ANOVA: Theory and Applications with R. Springer, New York

Battauz M, Bellio R, Gori E (2005) A multilevel measurement error model for value-added assessment in education. In: Atti Convegno S.Co. 15–17 settembre 2005, Bressanone, pp. 91–96

Béguin AA, Glas CAW (2001) MCMC estimation and some model-fit analysis of multidimensional IRT models. Psychometrika 66(4):541–562

Benjamini Y, Hochberg Y (1995) Controlling the false discovery rate: a new and powerful approach to multiple testing. Journal of the Royal Statistical Society Series B 57:1289–1300

Berni R, Gonnelli C (2006) Planning and optimization of a numerical control machine in a multiple response case. Quality Reliability Engineering International 22:517–526

Bertaccini B, Bini M (2007) Valutazione del processo di formazione universitaria: un'analisi robusta degli abbandoni. In: Effectiveness of University Education in Italy: Employability, Competences, Human Capital. Physica-Verlag Springer, Heidelberg

Bertaccini B, Polverini F (2006) Automatic forward search. Proceedings of the Joint Statistical Meetings of the American Statistical Association, 06–10 August 2006, Seattle, MIRA: Digital Publishing, USA

Bertaccini B, Varriale R (2007) Robust analysis of variance: an approach based on the forward search. Computational Statistics & Data Analysis 51:5172–5183

Bertaccini B, Varriale R (2008) Robust random effect models: an approach based on the Forward Search (Working Paper No. 12), Dipartimento di Statistica "G. Parenti", Università degli Studi di Firenze, Firenze

Bhat CR (2001) Quasi-random maximum simulated likelihood estimation of the mixed multinomial logit model. Transportation Research: Part B: Methodological 35:677–693

Bianconcini S, Cagnone S, Mignani S, Monari P (2007) A latent curve analysis of unobserved heterogeneity in university student achievements. Statistica LV:40–56

Biggeri L, Bini M (2003) Performance evaluation of the university system in italy: a robust clustering approach to validate homogeneous groups of units. Proceedings of the Joint Statistical Meetings of the American Statistical Association, 03–07 August 2003, San Francisco MIRA: Digital Publishing, USA

Biggeri L, Bini M, Grilli L (2001) The transition from university to work: a multilevel approach to the analysis of the time to obtain the first job. Journal of the Royal Statistical Society A 164:293–305

Bini M (2004a) Robust multivariate methods for the analysis of the university performance. In: Studies in Classification, Data Analysis, and Knowledge Organization, Springer, Berlin

References

Bini M (2004b) Valutazione del processo di formazione universitaria: un'analisi robusta degli abbandoni. In: Aureli Cutillo, E. (Eds.), Strategie metodologiche per lo studio della transizione Università-lavoro. Cleup, Padova

Bini M, Bertaccini B (2004) Forward search nell'analisi di regressione. In: Aureli Cutillo, E. (Eds.), "Strategie metodologiche per lo studio della transizione Università-lavoro". Cleup, Padova

Bini M, Bertaccini B (2007) Evaluating the university educational process.a robust approach to the drop-out. In: Fabbris, L. (Ed.), Effectiveness of University Education in Italy: Employability, Competences, Human Capital, Physica-Verlag Springer, Heidelberg, pp. 55–69

Bini M, Bertaccini B, Petrucci A (2002) Robust diagnostics in regression models for the evaluation of the university teaching effectiveness: the case of graduates of florence. Proceedings of the Joint Statistical Meetings of the American Statistical Association, 10–16 August 2002, New York, MIRA: Digital Publishing, USA

Bini M, Bertaccini B, Polverini F (2003) The use of outliers for the evaluation of public policy activities: the first year college drop out rate in florence. Proceedings of the Joint Statistical Meetings of the American Statistical Association, 03–07 August 2003, San Francisco, MIRA: Digital Publishing, USA, pp. 1–8

Birch MW (1964) The detection of partial association i: the 2×2 case. Journal of the Royal Statistical Society, Series B 26:313–324

Bird SM (2004) Editorial: Performance monitoring in the public services. Journal of the Royal Statistical Society A 167:381–383

Bird SM, Cox D, Farewell VT, Goldstein H, Holt T, Smith PC (2005) Performance indicators: good, bad, and ugly. Journal of the Royal Statistical Society A 168:1–27

Birnbaum A (1968) Some latent trait models and their use in inferring an examinee's ability. In: Lord, F.M. and Novick, M.R. (Eds.), Statistical Theories of Mental Test Scores, Addison-Wesley, Reading, MA, pp. 397–472

Bishop YMM, Fienberg SE, Holland PW (1975) Discrete Multivariate Analysis. The MIT Press, Cambridge, MA

Bock RD, Aitkin M (1981) Marginal maximum likelihood estimation of item parameters: application of an EM algorithm. Psychometrika 46(4):443–459

Boero G, Staffolani S (Eds.) (2006) Performance accademica e tassi di abbandono. Un'analisi dei primi effetti della riforma universitaria. CUEC, Cagliari

Bollen KA (1989) Structural Equations with Latent Variables. John Wiley and Sons, New York

Bollen KA, Curran PJ (2006) Latent Curve Models: A Structural Equation Perspective. John Wiley and Sons, New York.

Bolzan M, Marson M, Selle P, Verza N (2008) Una mappa concettuale per la rilevazione delle competenze dei dirigenti degli enti locali. In: Convegno OUTCOMES 2: Modelli e metodi per abbinare profili formativi e bisogni di professionalità di comparti del terziario avanzato, Università di Padova, Padova

Bonnini S, Corain L, Salmaso L (2006) A new statistical procedure to support industrial research into new product development. Quality and Reliability Engineering International 22(5):555–566

Box GEP, Andersen SL (1955) Permutation theory in the derivation of robust criteria and the study of departures from assumption. Journal of the Royal Statistical Society Series B 17:1–26

Box GEP, Hunter JS, Hunter WS (1978) Statistics for Experimenters. John Wiley and Sons, New York

Boxall PC, Adamowicz WL (2002) Understanding heterogeneous preferences in random utility models: a latent class approach. Environmental and Resource Economics 23:421–446

Bradlow ET, Hu Y, Ho TH (2004) A learning-based model for imputing missing levels in partial conjoint profiles. Journal of Marketing Research 41:369–381

Braun H, Wainer H (2007) Value-added modeling. In: Rao, C.R. and Sinharay, S. (Eds.), Handbook of Statistics 26: Psychometrics, Elsevier, Amsterdam, pp. 475–501

Browne MW (1993) Structured latent curve models. In: Cuadras, C.M. and Rao C.R. (Eds.), Multivariate Analysis: Future Directions. Elsevier Science, New York, pp. 171–197.

Browne MW, Du Toit SHC (1991) Models for learning data. In: Collins, L.M. and Horn, J.L. (Eds.), Best Methods for the Analysis of Change. American Psychological Association, Washington, DC, pp. 47–68

Browne W, Goldstein H, Rasbash J (2001) Multiple membership multiple classification (MMMC) models. Statistical Modelling 1:103–124

Burgess L, Street DJ (2005) Optimal designs for choice experiments with asymmetric attributes. Journal of Statistical Planning and Inference 134:288–301

Cagnone S, Mignani S (2007) Assessing the goodness of fit of la latent variable model for ordinal data. Metron 24:337–361

Cagnone C, Mignani S, Gardini A (2004) New developments of latent variable models with ordinal data. In: Proceedings of the XLI Scientific meeting of the Italian Statistical Society, Padova: Cleup, pp. 221–231

Cagnone S, Moustaki I, Vasdeskis V (2009) Latent variable models for multivariate longitudinal ordinal responses. British Journal of Mathematical and Statistical Psychology 62(2): 401–415

Campbell D, Hutchinson GW, Scarpa R (2008) Incorporating discontinuous preferences into the analysis of discrete choice experiments. Environmental and Resource Economics 41:401–417

Capursi V, Ghellini G (Eds.) (2008) Dottor Divago. Discernere, valutare e governare la nuova università. Franco Angeli, Milano

Chen YI, Wolfe DA (1990) A study of distribution-free tests for umbrella alternatives. Biometrical Journal 32:47–57

Chiandotto B (2004) Sulla misura della qualitá della formazione universitaria. Studi e note di Economia 3:27–61

Chiandotto B, Giusti C (2006) Gli effetti della riforma universitaria sui tempi di conseguimento del titolo. In: Crocetta, C. (Eds.), Metodi e modelli per la valutazione del sistema universitario, Vol. 9, Cleup, Padova

Chiandotto B, Varriale R (2006) La valutazione dei servizi di supporto alla didattica. In: Crocetta, C. (Eds.), Metodi e modelli per la valutazione del sistema universitario, Vol. 9, Cleup, Padova

Chiandotto B, Bertaccini B, Varriale R (in press) The effectiveness of university education: a structural equation model. In: Arabie, P. (Eds.), Studies in Classification, Data Analysis and Knowledge Organization: Data Analysis, Machine Learning and Applications Studies. Springer, New York

Chiandotto B, Grilli L, Rampichini C (Eds.) (2005) Valutazione dei processi formativi di terzo livello: contributi metodologici. No. 12 in Collana Valmon, Università degli studi di Firenze, Firenze, URL http://valmon.ds.unifi.it

Corain L, Salmaso L (2004) Multivariate and multistrata nonparametric tests: the npc method. Journal of Modern Applied Statistical Methods 3:443–461

Corain L, Salmaso L (2007) A nonparametric method for defining a global preference ranking of industrial products. Journal of Applied Statistics 34(2):203–216

Corduas M (2008a) Clustering CUB models by Kullback-Liebler divergence. In: Joint SFC-CLADAG Meeting Proceedings, ESI, pp. 245–248

Corduas M (2008b) A statistical procedure for clustering ordinal data. Quaderni di Statistica 10:177–189

Corduas M (2008c) A study on university students' opinions about teaching quality: a model based approach for clustering data. In: Proceedings of DIVAGO Meeting, University of Palermo, 10–12 July 2008

Corduas M (2008d) A testing procedure for clustering ordinal data by CUB models. In: Proceedings of the Joint SFC-CLADAG Meeting, ESI, Naples, pp. 245–248

Currim IS (1981) Using segmentation approaches for better prediction and understanding from consumer mode choice models. Journal of Marketing Research 18:301–309

Danaher PJ (1997) Using conjoint analysis to determine the relative importance of service attributes measured in customer satisfaction surveys. Journal of Retailing 73:235–260

De Bruyn A, Liechty JC, Huizingh EKRE, Lilien GL (2008) Offering online recommendations with minimum customer input through conjoint-based decision aids. Marketing Science 27:443–470

References

De Leeuw J, Meijer E (Eds.) (2008) Handbook of Multilevel Analysis. Springer, New York

D'Elia A (2000) Il meccanismo dei confronti appaiati nella modellistica per graduatorie: sviluppi statistici ed aspetti critici. Quaderni di Statistica 2:173–416

D'Elia A (2001) Efficacia didattica e strutture universitarie: la valutazione mediante un approccio modellistico. In: Atti Convegno SIS su: "Processi e metodi statistici di valutazione", ECRI, Roma, pp. 21–24

D'Elia A (2008) A statistical modelling approach for the analysis of TMD chronic pain data. Statistical Methods in Medical Research 17:389–403

D'Elia A, Piccolo D (2005) A mixture model for preference data analysis. Computational Statistics & Data Analysis 49:917–934

Dellaert BGC, Brazell JD, Louviere JJ (1999) The effect of attribute variation on consumer choice consistency. Marketing Letters 10:139–147

Dempster AP, Laird NM, Rubin DB (1977) Maximum likelihood from incomplete data via the *EM* algorithm (with Discussion). Journal of the Royal Statistical Society, Series B 39:1–38

DeShazo JR, Fermo G (2002) Designing choice sets for stated preference methods: the effects of complexity on choice consistency. Journal of Environmental Economics and Management 44:123–143

Diggle PJ, Liang KY, Zeger SL (1994) Analysis of Longitudinal Data. Clarendon press, Oxford

Donoho DL, Huber PJ (1983) The notation of breakdown point. In A festschrift for Erich Lehmann. edited by Bickel, P. and Doksum, K. and Hodges, J. L., Wadsworth, Belmont, CA.

Draper D, Gittoes M (2004) Statistical analysis of performance indicators in uk higher education. Journal of the Royal Statistical Society A 167:449–474

Du X, Jiao RJ, Tseng MM (2006) Understanding customer satisfaction in product customization. The International Journal of Advanced Manufacturing Technology 31:396–406

Du Toit SHC, Cudeck R (2001) The analysis of nonlinear random coefficient regression models with lisrel using constraints. In: Cudeck, R. du Toit, S. and Sörbom D. (Eds.), Structural Equation Modeling: Present and Future. Scientific Software I: Lincolnwood, IL, pp. 259–278

Dunson BD (2003) Dynamic latent trait models for multidimensional longitudinal data. Journal of the American Statistical Association 98(463):555–563

Edgington E, Onghena P (2007) Randomization Tests. 4th edn. Chapman and Hall, London

Everitt BS (1988) A Monte Carlo investigation of the likelihood ratio test for number of classes in latent class analysis. Multivariate Behavioral Research 23:531–538

Fabbri D, Fazioli R, Filippini M (1996) L'intervento pubblico e l'efficienza possibile. Il Mulino, Bologna

Fabbris L (Ed.) (2007) Effectiveness of University Education in Italy: Employability, Competences, Human Capital. Physica-Verlag, Heidelberg

Fawcett RF, Salter KC (1984) A monte carlo study of the f test and three tests based on ranks of treatment effects in randomized block designs. Communications in Statistics B-13:213–225

Fayers PM, Hand DJ (2002) Casual variables, indicator variables and measurement scales: an example from quality of life. Journal of the Royal Statistical Society Series A 165:233–261

Ferrão ME, Goldstein H (2009) Adjusting for measurement error in the value added model: evidence from portugal. Quality and Quantity (forthcoming) DOI 101007/s11135-008-9171-1

Finos L, Salmaso L (2007) Fdr- and fwe-controlling methods using data-driven weights. Journal of Statistical Planning and Inference 137:3859–3870

Fischer GW, Luce MF, Jia J (2000) Attribute conflict and preference uncertainty: effects on judgment time and error. Management Science 46:88–103

Fox JP, Glas CAW (2001) Bayesian estimation of a multilevel IRT model using Gibbs sampling. Psychometrika 66(2):271–288

Fox JP, Klein-Entink R, van der Linden WJ (2007) Modeling of response times with the package cirt. Journal of Statistical Software 20(7):1–14

Friedman M (1937) The use of ranks to avoid the assumption of normality implicit in the analysis of variance. Journal of the American Statistical Association 32:675–701

Fuller WA (1987) Measurement Error Models. John Wiley and Sons, New York

Garrod GD, Scarpa R, Willis KG (2002) Estimating the benefits of traffic calming on through routes: a choice experiment approach. Journal of Transport Economics and Policy 36:211–231

Gelfand AE, Smith AFM (1990) Sampling-based approaches to calculating marginal densities. Journal of the American Statistical Association 85:398–409

Geman S, Geman D (1984) Stochastic relaxation, Gibbs distributions and the Bayesian restoration of images. IEEE Transactions on Pattern Analysis and Machine Intelligence 6:721–741

Gilbride TJ, Allenby GM (2004) A choice with conjunctive disjunctive and compensatory screening rules. Marketing Science 23:391–406

Giusti C, Varriale R (2008) Un modello multilivello per l'analisi longitudinale della valutazione della didattica. In: Di Maio, A., Gallo, M. and Simonetti, B. (Eds.), Metodi, Modelli e Tecnologie dell'informazione a Supporto delle Decisioni - second part: Applicazioni, Franco Angeli - Pubblicazioni DASES, Milano, pp. 122–129

Goldstein H (2003) Multilevel Statistical Models, 3rd edn. Arnold, London

Goldstein H, Healy MJR (1995) The graphical presentation of a collection of means. Journal of the Royal Statistical Society A 158:175–177

Goldstein H, Leckie G (2008) School league tables: what can they really tell us? Significance 5:67–69

Goldstein H, McDonald RP (1988) A general model for the analysis of multilevel data. Psychometrika 53:455–467

Goldstein H, Spiegelhalter DJ (1996) League tables and their limitations: statistical issues in comparisons of institutional performances. Journal of the Royal Statistical Society A 159:385–443

Goldstein H, Burgess S, McConnell B (2007) Modelling the effect of pupil mobility on school differences in educational achievement. Journal of the Royal Statistical Society A 170:941–954

Goodman LA (1974) Exploratory latent structure analysis using both identifiable and unidentifiable models. Biometrika 61:215–231

Gori E, Vittadini G (1999) La valutazione dell'efficienza ed efficacia dei servizi alla persona. impostazione e metodi. In: Gori, E., and Vittadini, G. (Eds.), Qualità e valutazione nei servizi di pubblica utilità, Etas, RCS Libri, Milano, pp. 121–241

Gottard A, Grilli L, Rampichini C (2007) A chain graph multilevel model for the analysis of graduates' employment. In: L. Fabbris (Ed.), Effectiveness of University Education in Italy: Employability, Competences, Human Capital. Physica-Verlag, Heidelberg, pp. 169–182

Green PE (1984) Hybrid models for conjoint analysis: an expository review. Journal of Marketing Research 21:155–169

Green PE, Rao V (1971) A measurement for quantifying judgement data. Journal of Marketing Research 8:355–363

Green PE, Srinivasan V (1978) Conjoint analysis in consumer research issues and outlook. Journal of Marketing Research 5:103–123

Green PE, Srinivasan V (1990) Conjoint analysis in marketing: new developments with implications for research and practice. Journal of Marketing 54:3–19

Green PE, Goldberg SM, Montemayor M (1981) A hybrid utility estimation model for conjoint analysis. Journal of Marketing 45:33–41

Greene WH, Hensher DA (2003) A latent class model for discrete choice analysis: contrasts with mixed logit. Transportation Research: Part B: Methodological 37:681–698

Grilli L (2005) The random effects proportional hazards model with grouped survival data: a comparison between the grouped continuous and continuation ratio versions. Journal of the Royal Statistical Society A 168:83–94

Grilli L, Rampichini C (2007a) Multilevel factor models for ordinal variables. Structural Equation Modeling 14:1–25

Grilli L, Rampichini C (2007b) A multilevel multinomial logit model for the analysis of graduates' skills. Statistical Methods and Applications 16:381–393

Grilli L, Rampichini C (2007c) Selection bias in linear mixed models. Working Papers 2007/10 Dipartimento di Statistica 'G Parenti', Università di Firenze

Groggel DJ (1987) A monte carlo study of rank tests for block designs. Communication in Statistics – Simulation and Computation 16:601–620

Hagerty MR (1985) Improving the predictive power of conjoint analysis: the use of factor analysis and cluster analysis. Journal of Marketing Research 22:168–184

Haley DC (1952) Estimation of the dosage mortality relationship when the dose is subject to error. Technical Report 15 (Office of Naval Research Contract No. 25140, NR-342-022), Stanford University: Applied Mathematics and Statistics Laboratory

Hanley N, Mourato S, Wright RE (2001) Choice modelling approaches: a superior alternative for environmental evaluation? Journal of Economic Surveys 15:435–462

Hanushek EA (1986) The economics of schooling: production and efficiency in public schools. Journal of Economic Literature 24:1141–1177

Hartlaub BA, Wolfe DA (1999) Distribution-free ranked-set sample procedures for umbrella alternatives in the m-sample setting. Environmental and Ecological Statistics 6:105–118

Hartman RS, Doane MJ, Woo CK (1991) Consumer rationality and the status quo. The Quarterly Journal of Economics 106:141–162

Herriges JA, Phaneuf DJ (2002) Inducing patterns of correlation and substitution in repeated logit models of recreation demand. American Journal of Agricultural Econonomics 84:1076–1090

Hettmansperger TP, Norton RM (1987) Tests for patterned alternatives in k-sample problems. Journal of the American Statistical Association 82:292–299

Hirotsu C (1998) Isotonic inference. In: Encyclopedia of Biostatistics, John Wiley and Sons, New York, pp. 2107–2115

Hoeffding W (1952) The large-sample power of tests based on permutations of observations. Annals of Mathematical Statistics 23:169–192

Hollander M, Wolfe DA (1999) Nonparametric Statistical Methods, 2nd edn. Wiley Series in Probability and Statistics, New York

Hollander M, Wolfe DA (2003) An Introduction to Multivariate Statistical Analysis. John Wiley and Sons, New York

Hong G, Raudenbush SW (2008) Causal inference for time-varying instructional treatments. Journal of Educational and Behavioral Statistics 33:333–362

Hox J (2002) Multilevel Analysis: Techniques and Applications. Quantitative Methodology Series, Lawrence Erlbaum Associate, London

Huber PJ (1981) Robust Statistics. John Wiley and Sons, New York.

Huber P, Ronchetti E, Feser V (2004) Estimation of generalized linear latent variable models. Journal of the Royal Statistical Society, Series B 66:893–908

Hynes S, Hanley N, Scarpa R (2008) Effects on welfare measures of alternative means of accounting for preference heterogeneity in recreational demand models. American Journal of Agricultural Economics 90:1011–1027

Iannario M (2007) A statistical approach for modelling Urban Audit Perception Surveys. Quaderni di Statistica 9:149–172

Iannario M (2008a) A class of models for ordinal variables with covariates effects. Quaderni di Statistica 10:53–72

Iannario M (2008b) Dummy variables in CUB models. Statistica LXVIII, 179–200

Iannario M (2009a) A comparison of preliminary estimators in a class of ordinal data models. Statistica & Applicazioni VII, 25–44

Iannario M (2009b) Modelling *shelter* choices in ordinal data surveys, Department of Statistical Sciences, University of Naples Federico II, preliminary report, submitted

Iannario M (2009c) A note on the identifiability of a mixture model for ordinal data, submitted

Iannario M, Piccolo D (2008) Modelli statistici per l'analisi dei processi decisionali. In: Iannaccone, A., and Storti, G. (Eds.), Decision Making. Un approccio interdisciplinare, University of Salerno, pp 41–61

Iannario M, Piccolo D (2009) A program in R for CUB models inference. Available via Internet, URL http://www.dipstat.unina.it/CUBmodels/, Version 2.0

Jiao J, Tseng MM (2004) Customizability analysis in design for mass customization. Computer-Aided Design 36:745–757

Jiao RJ, Xu Q, Du J, Zhang Y, Helander M, Khalid HM, Helo P, Ni C (2007) Analytical affective design with ambient intelligence for mass customization and personalization. International Journal of Flexible Manufacturing Systems 19:570–595

Jin H, Rubin DB (2009) Public schools versus private schools: causal inference with partial compliance. Journal of Educational and Behavioral Statistics 34:24–45

Johnson MR (1974) Trade-off analysis of consumer values. Journal of Marketing Research 11:121–127

Jöreskog K (1990) New developments in lisrel: analysis of ordinal variables using polychoric correlations and weighted least squares. Quality and Quantity 24:387–404

Jöreskog K (2002) Structural equation modeling with ordinal variables using. In: Scientif Software International, Chicago

Jöreskog K, Moustaki I (2001) Factor analysis of ordinal variables: a comparison of three approaches. Multivariate Behavioral Research 36:347–387

Jöreskog K, Sörbom D (1988) LISREL 8: Users' Reference Guide. 2nd edn. SSI, Chicago

Kazemzadeh RB, Behzadian M, Aghdasi M, Albadvi A (2008) Integration of marketing research techniques into house of quality and product family design. The International Journal of Advanced Manufacturing Technology 8:1–15, URL www.springerlink.com/content/102823/?k=kazemzadeh

Kessels R, Goos P, Vanderbroek M (2004) Comparing algorithms and criteria for designing bayesian conjoint choice experiments. Research Report, Department of Applied Economics, Katholieke Universiteit Leveun, Belgium 427:1–38

Khuri AI, Cornell JA (1987) Response surfaces: designs and analyses. Dekker, New York

Kim JS, Frees EW (2007) Multilevel modeling with correlated effects. Psychometrika 72:505–533

King G, Tomz M, Wittenberg J (2000) Making the most of statistical analyses: improving interpretation and presentation. American Journal of Political Science 44:341–355

Koheler KJ, Larntz K (1980) An empirical investiogation of goodness-of-fit statistics for sparse multinomials. Journal of the American Statistical Association 75:336–244

Kössler W (2006) Some c-sample rank tests of homogeneity against umbrella alternatives with unknown peak. Journal of statistical computation and simulation 76:57–74

Kullback S (1959) Information Theory and Statistics. Dover, New York

Ladd HF, Walsh RP (2002) Implementing value-added measures of school effectiveness: getting the incentives right. Economics of Education Review 21:1–17

Lago A, Pesarin F (2000) Non parametric combination of dependent rankings with application to the quality assessment of industrial products. Metron LVIII:39–52

Laird NM, Ware JH (1982) Random effects models for longitudinal data. Biometrics 38:963–974.

Lamond E (1970) Methods for Sensory Evaluation of Food. Department of Agriculture, Otawa, Canada

Lawrence L, Marsh C (2004) The econometrics of higher education: editor's view. Journal of Econometrics 121:1–18

Lazarsfield PF, Henry NW (1968) Latent Structure Analysis. Houghton Mifflin, Boston

Leckie G (2009) The complexity of school and neighbourhood effects and movements of pupils on school differences in models of educational achievement. Journal of the Royal Statistical Society A 172:537–554

Leckie G, Goldstein H (2009) The limitations of using school league tables to inform school choice. The Centre for Market and Public Organization 09/208, Department of Economics, University of Bristol, UK

Lehmann EL, D'Abrera HJM (2006) Nonparametrics: Statistical Methods Based on Ranks. Springer, Berlin

Lenk PJ, DeSarbo WS, Green PE, Young MR (1996) Hierarchical Bayes conjoint analysis: recovery of partworth heterogeneity from reduced experimental designs. Marketing Science 15:173–191

Leti G (1979) Distanze e indici statistici. La Goliardica editrice, Roma

Longford N, Muthén BO (1992) Factor analysis for clustered observations. Psychometrika 57:581–597

Lord FM (1952) A theory of test scores. Psychometric Monograph No. 7

Lord FM, Novick MR (1968) Statistical Theories of Mental Test Scores. Addison-Wesley, Reading, MA

Lubke GH, Muthén BO (2005) Investigating population heterogeneity with factor mixture models. Psychological Methods 10:21–39

Lukočiené O, Vermunt JK (2004) Determining the number of components in mixture models for hierarchical data. In: Fink, A., Berthold, L., Seidel, W. and Ultsch, A. (Eds.), Advances in data analysis, data handling and business intelligence. Springer, Berlin-Heidelberg

Mack GA, Wolfe DA (1981) K-sample rank tests for umbrella alternatives. Journal of the American Statistical Association 76:175–181

Magel RC, Qin L (2003) A non-parametric test for umbrella alternatives based on ranked-set sampling. Journal of Applied Statistics 30:925–937

Magidson J, Vermunt JK (2001) Latent class factor and cluster models, bi-plots and related graphical displays. Sociological Methodology 31:223–264

Manly BFJ (1997) Randomization, Bootstrap and Monte Carlo Methods in Biology. 2nd edn. Chapman and Hall, London

Marcus R, Peritz E, Gabriel KR (1976) On closed testing procedures with special reference to ordered analysis of variance. Biometrika 63:655–660

Martini A, Ricci R (2007) PISA2003 mathematical performance of italian students: multilevel analysis for each secondary school category. Induzioni 34:25–46

Matteucci M (2007a) Item response theory models for the competence evaluation: towards a multidimensional approach in the University guidance. PhD thesis, Statistics Department, University of Bologna (Italy)

Matteucci M (2007b) A multidimensional item response theory approach for the University guidance. In: Eum Edizioni Università di Macerata (Eds.), Book of short papers, Sixth Scientific Meeting of the CLAssification and Data Analysis Group of the Italian Statistical Society, Macerata, pp. 697–700

Matteucci M, Stracqualursi L (2006) Student assessment via graded response model. Statistica 4:435–447

Matteucci M, Veldkamp BP (2008) Including empirical prior information in test administration. In: Edizioni Scientifiche Italiane (Eds.), Book of short papers, First Joint Meeting of the Société Francophone de Classification and the CLAssification and Data Analysis Group of SIS, Caserta, pp. 97–100

Matteucci M, Mignani S, Ricci R (2008) Self-evaluation test for student guidance: the case of the University of Bologna. On-line paper, URL http://amsacta.cib.unibo.it/archive/00002431/

Maydeu-Olivares A, Harry J (2005) Limited- and full-information estimation and goodness-of-fit testing in 2^n contingency tables: A unified framework. Journal of American Statistical Association 100:1009–1020

McArdle JJ (1988) Dynamic but structural equation modeling of repeated measures data. In: Nesselroade, J.R. and Cattell, R.B. (Eds.), The Handbook of Multivariate Experimental Psychology. New York: Plenum Press I:561–614

McCormick WT, Schweitzer PJ, White TW (1972) Problem decomposition and data reorganization by a clustering tecnique. Operation Research 20:993–1009

McCullagh P, Nelder JA (1983) Generalized Linear Models. Chapman and Hall, New York

McFadden D (1974) Conditional logit analysis of qualitative choice behaviour. In: Zabremska (Eds.), Frontiers of Econometrics. Academic Press, New York

McFadden D, Train K (2000) Mixed MNL for discrete response. Journal of Applied Econometrics 15:447–450

McLachlan GJ, Peel D (2000) Finite Mixture Models. John Wiley and Sons, New York

Meilgaard M, Civille GV, Carr BT (2006) Sensory Evaluation Techniques. 4th edn. CRC Press, Boca Raton, Florida

Meredith W (1993) Measurement invariance, factor analysis and factorial invariance. Psycometrika 58:525–543

Meredith W, Tisak J (1990) Latent curve analysis. Psychometrika 55:107–122

Michalek JJ, Feinberg FM, Papalambros PY (2005) Linking marketing and engineering product design decisions via analytical target cascading. The Journal of Product Innovation Management 22:42–62

Millen BA, Wolfe DA (2005) A class of nonparametric tests for umbrella alternatives. Journal of Statistical Research 39:7–24

Montgomery DC (2005) Design and Analysis of Experiments. 6th edn. John Wiley and Sons, New York

Moore WL (1980) Levels of aggregation in conjoint analysis: an empirical comparison. Journal of Marketing Research 17:516–523

Moustaki I (2000) A latent variable model for ordinal data. Applied psychological measurement 24:211–223

Moustaki I (2003) A general class of latent variable models for ordinal manifest variables with covariates effects on the manifest and latent variables. British Journal of Mathematical and Statistical Psychology 56:337–357

Moustaki I, Knott M (2000) Generalized latent trait models. Psychometrika 65:391–411

Muthén BO (1984) A general structural equation model with dichotomous ordered categorical and continuous latent variable indicators. Psychometrika 49:115–132

Muthén BO (1989) Latent variable modeling in heterogeneous populations. Psycometrika 54:115–132

Muthén BO (2004) Latent variable analysis: growth mixture modeling and related techniques for longitudinal data. In: Kaplan, D. (Eds.), Handbook of Quantitative Methodology for the Social Sciences, Sage Publications. Newbury Park, CA, pp. 345–368

Muthén BO (2008) Latent variable hybrids: Overview of old and new models. In: Hancock, G.R. and Samuelsen, K.M. (Eds.), Advances in Latent Variable Mixture Models. Information Age Publishing, Inc., Charlotte, NC

Muthén BO, Asparouhov T (2008) Growth mixture modeling: analysis with non-gaussian random effects. In: Fitzmaurice, G., Davidian, M., Verbeke, G. and Molenberghs, G. (Eds.), Longitudinal Data Analysis. Chapman & Hall/CRC Press, Boca Raton, FL

Muthén BO, Khoo ST (1998) Longitudinal studies of achievement growth using latent variable modeling. Learning and Individual Differences 10:73–101

Muthén BO, Muthén L (1998–2007) Mplus User's Guide. 5th edn. Technical Report, Muthén and Muthén, Los Angeles

Myers RH, Montgomery DC (1995) Response Surface Methodology. John Wiley and Sons, New York

Nardo M, Saisana M, Saltelli A, Tarantola S, Hoffman A, Giovannini E (2008) Handbook on Constructing Composite Indicators: Methodology and User Guide. OECD Publishing, Paris

Netzer O, Srinivasan V (2007) Adaptive self-explication of multi-attribute preferences. Research Paper, Stanford University Graduate School of Business 1979:1–42, URL https://gsbapps.stanford.edu/researchpapers/

Netzer O, Toubia O, Bradlow ET, Dahan E, Evgeniu T, Feinberg FM, Feit EM, Hui SK, Johnson J, Liechty JC, Orlin JB, Rao VR (2008) Beyond conjoint analysis: advances in preference measurement. Marketing Letters 19:337–354

Neuhäuser M, Leisler B, Hothorn LA (2003) A trend test for the analysis of multiple paternity. Journal of Agricultural, Biological & Environmental Statistics 8:29–35

Novick MR (1966) The axioms and principal results of classical test theory. Journal of Mathematical Psychology 3(1):1–18

Nylund KL, Muthén BO, Asparouhov T (2007) Deciding on the number of classes in latent class analysis and growth mixture modeling: a Monte Carlo simulation study. Structural Equation Modeling: A Multidisciplinary Journal 14:535–569

O'Gorman TW (2001) A comparison of the f-test, friedman's test, and several aligned rank tests for the analysis of randomized complete blocks. Journal of Agricultural, Biological, and Environmental Statistics 6:367–378

Palardy G, Vermunt JK (2009) Multilevel growth mixture models for classifying group-level observations, Journal of Educational and Behavioral Statistics

Pan G (1996) Distribution-free tests for umbrella alternatives. Communications in Statistics – Theory and Methods 25:3185–3194
Pan G (2008) Distribution-free confidence procedure for umbrella orderings. Australian & New Zealand Journal of Statistics 38:161–172
Pawitan Y (2001) In All Likelihood: Statistical Modelling and Inference Using Likelihood. Clarendon Press, Oxford
Pesarin F (2001) Multivariate Permutation Tests with Applications in Biostatistics. John Wiley and Sons, Chichester
Pesarin F (2002) Extending permutation conditional inferences to unconditional ones. Statistical Methods & Applications 11:161–173
Piccolo D (2003) On the moments of a mixture of uniform and shifted binomial random variables. Quaderni di Statistica 5:85–104
Piccolo D (2006) Observed information matrix for MUB models. Quaderni di Statistica 8:33–78
Piccolo D, D'Elia A (2008) A new approach for modelling consumers' preferences. Food Quality and Preference 19:247–259
Punj G, Stewart DW (1983) Cluster analysis in marketing research: review and suggestions for application. Journal of Marketing Research 20:134–148
Quade D (1979) Using weighted rankings in the analysis of complete blocks with additive block effects. Journal of the American Statistical Association 74:680–683
Rabe-Hesketh S, Skrondal A, Pickels A (1996) Generalized multilevel structural equation modeling. Psychometrika 69:167–190
Rasbash J, Goldstein H (1994) Efficient analysis of mixed hierarchical and cross-classified random structures using a multilevel model. Journal of Educational and Behavioral Statistics 19:337–350
Rasch G (1960) Probabilistic Models for Some Intelligence and Attainment Tests. Danish Institute for Educational Research, Copenhagen
Raudenbush SW (1993) A crossed random effects model for unbalanced data with applications in cross-sectional and longitudinal research. Journal of Educational Statistics 18:321–349
Raudenbush SW (2004) What are value-added models estimating and what does this imply for statistical practice? Journal of Educational and Behavioral Statistics 21:121–129
Raudenbush SW, Bryk AS (2002) Hierarchical Linear Models. 2nd edn. Sage Publications, Thousand Oaks
Raudenbush SW, Willms JD (1995) The estimation of school effects. Journal of Educational and Behavioral Statistics 20:307–335
Read TRC, Cressie NAC (1988) Goodness-of-fit Statistics for Discrete Multivariate Data. Springer-Verlag, New York
Reiser M (1996) Analysis of residual for the multinomial item response model. Psychometrika 61:509–528
Reiser M, Lin Y (1999) A goodness-of-fit test for the latent class model when expected frequencias are small. Sociological Methodology 39:81–111
Robinson TJ, Brenneman WA, Myers WR (2006) Process optimization via robust parameter design when categorical noise factors are present. Quality Reliability Engineering International 22:307–320
Rousseeuw PJ (1984) Least median of square regression. Journal of the American Statistical Association 85:633–639
Rousseeuw PJ, Leroy AM (1987) Robust Regression and Outlier Detection. John Wiley and Sons, New York
Roy J, Lin X (2000) Latent variable for longitudinal data with multiple continuous outcomes. Biometrics 56:1047–1054
Rubin DB, Stuart EA, Zanutto EL (2004) A potential outcomes view of value-added assessment in education. Journal of Educational and Behavioral Statistics 29:103–116
Sanders WL, Horn SP (1994). The Tennessee Value-Added Assessment System (TVAAS): Mixed methodology in educational assessment. Journal of Personnel Evaluation in Education, 8:299-311.

Sandor Z, Wedel M (2002) Profile construction in experimental choice designs for mixed logit models. Marketing Science 21:445–475

Sandor Z, Wedel M (2005) Heterogeneous conjoint choice designs. Journal of Marketing Research 42:210–218

Scarpa R, Thiene M (2005) Destination choice models for rock climbing in the Northeastern Alps:a latent-class approach based on intensity of preferences. Land Economics 81:426–444

Scarpa R, Ruto ESK, Kristjanson P, Radeny M, Druker AG, Rege JEO (2003) Valuing indigenous cattle breeds in Kenya: an empirical comparison of stated and revelaed preference value estimates. Ecological Economics 45:409–426

Scarpa R, Thiene M, Train K (2007) Utility in WTP space: a tool to address confounding random scale effects in destination choice to the Alps. Working Paper in Economics, Department of Economics, University of Waikato, New Zealand 15:1–22

Schilling S, Bock RD (2005) High-dimensional maximum marginal likelihood item factor analysis by adaptive quadrature. Psychometrika 70:533–555

Schütte S, Eklund J (2005) Design of rocker switches for work-vehicles – an application of kansei engineering. Applied Ergonomics 36:557–567

Schwarz G (1978) Estimating the dimention of a model. Annals of Statistics 6:461–464

Scott A (2002) Identifying and analysing dominant preferences in discrete choice experiments: an application in health care. Journal of Economic Psychology 23:383–398

Searle SR, Casella G, McCullogh CE (1992) Variance Components. John Wiley and Sons, New York

Segall DO (1996) Multidimensional adaptive testing. Psychometrika 61:331–354

Self SG, Liang KY (1987) Asymptotic properties of maximum likelihood estimators and likelihood ratio tests under nonstandard conditions. Journal of the American Statistical Association 82:605–610

Sepansky JH (2007) A modification on the friedman test statistic. Communication in Statistics – Simulation and Computation 36:783–790

Shi NZ (1998) Rank test statistics for umbrella alternatives. Communications in Statistics – Theory and Methods 17:2059–2073

Siegel S, Castellan N (1988) Nonparametric Statistics for the Behavioral Sciences. McGraw-Hill, New York

Simonoff JS (2003) Analyzing Categorical Data. Springer-Verlag, New York

Singer JD, Willett JB (2004) Applied Longitudinal Data Analysis: Modeling Change and Event Occurence. Oxford University Press, New York

Skrondal A, Rabe-Hesketh S (2004) Generalized Latent Variable Modeling: Multilevel, Longitudinal, and Structural Equation Models. Chapman and Hall/CRC Press, Boca Raton, FL

Snijders TAB, Berkhof J (2008) Diagnostic checks for multilevel models. In: de Leeuw, J. and Meijer, E. (Eds.), Handbook of Multilevel Analysis. Springer, New York

Snijders TAB, Bosker RJ (1999) Multilevel Analysis. An Introduction to Basic and Advanced Multilevel Modelling. Sage, London

Sonnier G, Ainslie A, Otter T (2007) Heterogeneity distributions of willingness-to-pay in choice models. Quantitative Marketing and Economics 5:313–331

Steele F, Vignoles A, Jenkins A (2007) The effect of school resources on pupil attainment: a multilevel simultaneous equation modelling approach. Journal of the Royal Statistical Society A 170:801–824

Strazzera E, Scarpa R, Calia P, Garrod GD, Willis KG (2003) Modelling zero values and protest responses in contingent valuation surveys. Applied Economics 35:133–138

Street DJ, Burgess L (2004) Optimal and near-optimal pairs for the estimation of effects in 2-level choice experiments. Journal of Statistical Planning and Inference 118:185–199

Swait J, Adamowicz W (2001) Choice environment, market complexity, and consumer behavior: a theoretical and empirical approach for incorporating decision compexity into models of consumer choice. Organizational Behavior and Human Decision Processes 86:141–167

Tekwe CD, Carter RL, Ma C, Algina J, Lucas ME, Roth J, Ariet M, Fisher T, Resnick MB (2004) An empirical comparison of statistical models for value-added assessment of school performance. Journal of Educational and Behavioral Statistics 29:11–36

Thissen D, Steinberg L (1986) A taxonomy of item response models. Psychometrika 51(4):567–577

Tollenar N, Mooljaart A (2003) Type I error and power of the parametric bootstrap goodness-of-fit test: Full and limited information. British Journal of Mathematical and Statistical Psychology 56:271–288

Toubia O, Hauser JR (2007) On managerially efficient experimental designs. Marketing Science 26:851–858

Toubia O, Hauser JR, Garcia R (2007) Probabilistic polyhedral methods for adaptive choice-based conjoint analysis. Theory and application. Marketing Science 26:596–610

Train KE (1998) Recreation demand models with taste differences over people. Land Economics 74:230–239

Tucker LR (1958) Determination of parameters of a functional relation by factor analysis. Psychometrika 23:19–23.

Tukey JW (1960) A survey of sampling from contaminated distribution. In: Olkin, I. (Ed.), Contributions to Probability and Statistics. University Press, Stanford, CA

van der Linden WJ (1999) Empirical initialization of the trait estimation in adaptive testing. Applied Psychological Measurement 23(1):21–29

van der Linden WJ (2005) Linear Models for Optimal Test Design. Springer-Verlag, New York

van der Linden WJ, Hambleton RK (1997) Handbook of Modern Item Response Theory. Springer-Verlag, New York

van Dijk B, Fok D, Paap R (2007) A rank-ordered logit model with unobserved heterogeneity in ranking capabilities. URL publishing.eur.nl/ir/repub/asset/8533/ei2007-07.pdf

Varriale R (2007) Multilevel Mixture Models for the analysis of the University Effectiveness. PhD thesis, University of Florence, Department of Statistics G. Parenti

Varriale R, Bertaccini B (2009) Robust random effects models: a diagnostic approach based on the forward search. submitted to Computational Statistics & Data Analysis

Veldkamp BP (1999) Multiple objective test assembly problems. Journal of Educational Measurement 36:253–266

Veldkamp BP, van der Linden WJ (2002) Multidimensional constrained adaptive testing. Psychometrika 67:575–588

Vermunt JK (2003) Multilevel latent class models. Sociological Methodology 33:213–239

Vermunt JK (2007) Multilevel mixture item response theory models: an application in education testing. Bulletin of the International Statistical Institute, 56th Session ISI 2007: Lisboa, Portugal 1253:1–4

Vermunt JK, Magidson J (2005) Factor analysis with categorical indicators: a comparison between traditional and latent class approaches. In: Van der Ark, A, Croon., M. and Sijtsma, K. (Eds.), New Developments in Categorical Data Analysis for the Social and Behavioral Sciences. Lawrence Erlbaum Associate Inc., Mahwah

Vitali O, Merlini A (1999) La qualità della vita: metodi e verifiche. Rivista Italiana di Demografia e Statistica 53:5–92

Wainer H (2004) Introduction to a special issue of the journal of educational and behavioral statistics on value-added assessment. Journal of Educational and Behavioral Statistics 29:1–3

Wainer H, Dorand NJ, Eignor D, Flaugher R, Green BF, Mislevy RJ, Steinberg L, Thissen D (1990) Computerized Adaptive Testing: A Primer. Lawrence Erlbaum Associates, Mahwah, NJ

Wang WC, Chen PH, Cheng YY (2004) Improving measurement precision of test batteries using multidimensional item response models. Psychological Methods 9(1):116–136

Wen CH, Koppelman FS (2001) A generalized nested logit model. Transportation Research: Part B: Methodological 35:627–641

Westfall PH, Tobias RD, Rom D, Wolfinger RD, Hochberg Y (1999) Multiple Comparisons and Multiple Tests using SAS. SAS Press, Cary, North Carolina

Wolfe DA (2006) Nonparametric distribution-free procedures for order restricted alternatives. In: Ahsanullah, M. and Raquab, M.Z. (Ed.), Recent Developments in Order Random Variables. Nova Science Publishers, New York

Wooldridge JM (2002) Econometric Analysis of Cross Section and Panel Data. The MIT Press, Cambridge, MA

Yu J, Goos P, Vandebroek ML (2009) Efficient conjoint choice designs in the presence of respondent heterogeneity. Marketing Science 28:122–135

Zwerina K, Huber J, Kuhfeld WF (1996) A general method for constructing efficient choice designs. Working paper-Fuqua School of Business-Duke University, Durham 27708:121–139, URL www.support.sas.com/techsup/technote/ts722e.pdf

Zwinderman AH (1991) A generalized rasch model for manifest predictors. Psychometrika 56(4):589–600

Index

agreement, 101
aligned rank test, 184, 185
alternative specific constant, 123

Bayesian framework, 33
Bayesian optimal designs , 124
bimodal distributions, 106
blocking, 181
Bologna process, 55

categorical data, 111
choice modelling, 119
choice-set, 121
choices' covariates, 103
chooser's covariates, 103
closed testing, 217–219
cluster analysis, 128
collateral information, 41
combining functions, 188, 191, 192, 201, 212, 216
completely latent model, 53
complexity of the design, 124
composite index, 163, 166
computerized adaptive testing, 32, 45
conditional maximum likelihood, 33
Conditional Monte Carlo (CMC) procedure, 187, 214–216
conjoint analysis, 119, 128
contingent ranking, 121
contingent valuation, 119, 120
CUB models, **102**
  clustering algorithm, 112
  EM algorithm, 106
  empirical evidence, 113–117
  expectation, 103
  extended, 105
    EM algorithm, 106–109
  fitting measures, 109
  GLM paradigm, 104
  identifiability, 103
  objects' covariates, 103
  observed information matrix, 109
  parameter estimation, 106
  satured, 110
  specification, 102
  stochastic component, 104
  subjects' covariates, 103
  systematic components, 104
  with covariates, 104

D-optimal design, 122, 124
data
  pooled, 11
    sufficiency of, 11
Data WareHouse, 55
diagnostic, 144, 145
  analysis, 142, 144
discrete Uniform, random variable, 102, 104
dissimilarity index, 109
distribution-free property, 12
dominant preferences, 124

educational assessment, 43
effectiveness, 62–65, 77, 142, 147
EM algorithm, 19
evaluation, 101
exact test, 11
exchangeability, 11, 186
  characterization of, 11
experimental design, 121, 130
exponential trajectory, 52

factor mixed model, 82
feeling, 101, 102

241

Index

finite-mixture model, 127
Forward Search, 146, 158, 159
fractional factorial design, 122
fuzziness, 101

Generalized Linear Latent Variable Model (GLLVM), 17, 50
Generalized Linear Models (GLM), 104
generalized nested logit, 123
Gibbs sampler, 34
global performance index, 161, 162
goodness-of-fit test, 23

heterogeneity, 120
hierarchical Bayes model, 122, 129
hybrid utility estimation model, 128

incomplete design, 40
independence of irrelevant alternatives , 125
invariance
    property, 12
isotonic inference, 195
item characteristic curve, 31
Item Response Theory (IRT), 30, 100, 117

joint maximum likelihood, 33

Kansei engineering , 129
kernel logit, 123
Kullback-Liebler divergence, 112

latent class model, 82, 123
latent curve model, 48, 49
latent variables, 99, 100, 104
league tables, 62, 65
likelihood ratio, 11
local independence, 30
longitudinal data, 24, 47

marginal maximum likelihood, 33
Markov Chain Monte Carlo, 34
Maximum Experimentwise Error Rate (MEER), 184, 218
measurement error, 76, 77
missing data analysis, 57
mixed multinomial logit, 121
model-based clustering, 111
multi-attribute methods, 119
multidimensionality, 37
multilevel modelling, 66, 69, 71, 72, 77, 87
    cross-classified, 72
    multiple membership, 72
multimodal distributions, 106
multinomial logit, 126

multiplicity, 183, 189, 190, 192, 216, 217, 219

non-stationary autoregressive model, 24
nonparametric combination, 188, 194, 196, 199, 212–216
NPC ranking, 212

optimization measures, 130, 132
ordinal observed variables, 17
orthogonal design, 122
outliers, 139, 140, 142–146, 152–158, 160

pairwise comparisons, 167, 186–190
panel data, 47
perception, 99, 100
performance, 103
permutation sample space, 11
permutation tests, 185, 213, 214
piecewise linear polynomial, 54
polynomial functions, 51
prior information, 41
product performances, 161
product quality evaluation, 161, 176
propensity, 103
pseudo-R^2, 110

random coefficients, 48
random effects, 24
random parameter logit, 126
random utility models, 121
randomness, 102
rank ordered logit, 123
rank transformation, 163, 183
ranking, 111, 121
rating, 110, 111, 121
RCB design, 181, 182, 185, 187
repeated choices, 123
residuals, 21
    empirical Bayes, 77, 78
    shrunken, 77
resolution, 122
response surface methodology, 120
revealed preference, 120
robust, 141–148, 158–160
    design, 130
    diagnostics, 139, 141, 143, 145, 147, 149, 151, 153, 155, 157, 159, 160
Robust Forward, 153, 156–158
robustness, 159

school effects, 69–71, 74, 77, 79
shelter choice, 104–106
shifted Binomial, random variable, 101, 102, 104

Index

simulation study, 173, 188, 190, 201
sparseness, 21
Spearman's rank correlation coefficient, 174, 176, 214
standardization and aggregation, 163, 171, 173, 176
stated preferences, 119
status-quo, 124, 131
stochastic ordering, 193, 196, 199
structural equation modelling, 49
students' profiles, 115

test
   distribution-free, 12
   invariant, 12
   nonparametric, 12

umbrella alternatives, 194, 195, 201, 203
uncertainty, 101–103
underlying variable approach, 49
unidimensionality, 30
university evaluation, 105, 112
university external effectiveness, 91
university orientation service, 112

value-added, 64, 67, 72, 73, 75, 76

willingness to pay, 119

CPSIA information can be obtained
at www.ICGtesting.com
Printed in the USA
LVHW051043030520
654914LV00002B/357